METHODS IN MOLECULAR BIOLOGY

Series Editor
John M. Walker
School of Life and Medical Sciences
University of Hertfordshire
Hatfield, Hertfordshire, UK

For further volumes:
http://www.springer.com/series/7651

For over 35 years, biological scientists have come to rely on the research protocols and methodologies in the critically acclaimed *Methods in Molecular Biology* series. The series was the first to introduce the step-by-step protocols approach that has become the standard in all biomedical protocol publishing. Each protocol is provided in readily-reproducible step-by-step fashion, opening with an introductory overview, a list of the materials and reagents needed to complete the experiment, and followed by a detailed procedure that is supported with a helpful notes section offering tips and tricks of the trade as well as troubleshooting advice. These hallmark features were introduced by series editor Dr. John Walker and constitute the key ingredient in each and every volume of the *Methods in Molecular Biology* series. Tested and trusted, comprehensive and reliable, all protocols from the series are indexed in PubMed.

Metabolic Gene Clusters

Methods and Protocols

Edited by

Hans-Wilhelm Nützmann

Biosciences, University of Exeter, Exeter, UK

Ramesha Thimmappa

Amity Institute of Genome Engineering, Amity University, Noida, Uttar Pradesh, India

Humana Press

Editors
Hans-Wilhelm Nützmann
Biosciences
University of Exeter
Exeter, UK

Ramesha Thimmappa
Amity Institute of Genome Engineering
Amity University
Noida, Uttar Pradesh, India

ISSN 1064-3745 ISSN 1940-6029 (electronic)
Methods in Molecular Biology
ISBN 978-1-0716-5383-8 ISBN 978-1-0716-5384-5 (eBook)
https://doi.org/10.1007/978-1-0716-5384-5

This Humana imprint is published by the registered company Springer Science+Business Media, LLC, part of Springer Nature.
The registered company address is: 1 New York Plaza, New York, NY 10004, U.S.A.

Preface

Metabolic gene clusters are defined as groups of adjacent genes that are together involved in the same metabolic pathway. Clusters predominantly consist of biosynthetic genes, and may also encode regulatory proteins, transporters and detoxifying enzymes. While the co-localisation of functionally related genes in clusters is the exception in eukaryotic genomes, metabolic gene clusters have been found in genomes across the eukaryotic tree of life. Over the past decades, metabolic gene clusters have served as treasure trove for the discovery of novel bioactive metabolites, have served as model systems to understand microbial genome evolution and eukaryotic gene regulation and have become targets for breeding of crops with improved quality. Here, we have assembled an array of methods that support research on metabolic gene clusters. The contributions range from animals to plants and microbes. They cover protocols on cluster identification, characterisation and evolutionary analyses, synthetic assembly and pathway re-constitution as well as cluster regulation and activation. We hope that this book will spur new discoveries in the field of metabolic gene clusters and bring the different areas of research closer together. We would like to highlight that all protocols can be applied beyond the topic of clustering in the fields of genome biology, metabolism and biotechnology.

At last, we would like to thank all authors who have contributed their time and expertise to this volume.

Exeter, UK — *Hans-Wilhelm Nützmann*
Noida, Uttar Pradesh, India — *Ramesha Thimmappa*

Contents

Contributors

ANDREW M. BAILEY • *School of Biological Sciences, Life Sciences Building, University of Bristol, Bristol, UK*
KARUNA BHASKARAN • *Amity Institute of Genome Engineering, Amity University, Noida, Uttar Pradesh, India*
LIANG DENG • *Key Laboratory of Combinatorial Biosynthesis and Drug Discovery (Ministry of Education), Wuhan University, Wuhan, China*
YANHONG GUO • *Key Laboratory of Saline-alkali Vegetation Ecology Restoration, Ministry of Education, College of Life Science, Northeast Forestry University, Harbin, China*
GEORGE JERVIS • *Biosciences, University of Exeter, Geoffrey Pope Building, Exeter, UK*
ZACHARY KONKEL • *Department of Plant Pathology, The Ohio State University, Columbus, OH, USA*
JENNY KOUKARA • *Laboratory of Plant and Environmental Biotechnology, Department of Biochemistry and Biotechnology, University of Thessaly, Larissa, Greece*
CHANG LI • *Department of Medicinal Chemistry and Natural Medicine Chemistry, College of Pharmacy, Harbin Medical University, Harbin, China*
DISHUAI LI • *College of Pharmacy, Nanjing University of Chinese Medicine, Nanjing, China; Jiangsu Collaborative Innovation Center of Chinese Medicinal Resources Industrialization, Nanjing University of Chinese Medicine, Nanjing, China*
CHANCHAN LIU • *College of Pharmacy, Nanjing University of Chinese Medicine, Nanjing, China; Jiangsu Collaborative Innovation Center of Chinese Medicinal Resources Industrialization, Nanjing University of Chinese Medicine, Nanjing, China*
LIN LIU • *Wuhan Frasergen Bioinformatics Co., Ltd., Wuhan, China*
ZHENHUA LIU • *Shanghai Collaborative Innovation Center of Agri-Seeds, School of Agriculture and Biology, Shanghai Jiao Tong University, Shanghai, China; Zhangjiang Institute for Advanced Study, Shanghai Jiao Tong University, Shanghai, China*
JIAOJIAO LU • *Key Laboratory of Saline-alkali Vegetation Ecology Restoration, Ministry of Education, College of Life Science, Northeast Forestry University, Harbin, China*
RAJESH CHANDRA MISRA • *Biochemistry and Metabolism Department, John Innes Centre, Norwich, UK; Plant Biotechnology Division, Council of Scientific and Industrial Research-Central Institute of Medicinal and Aromatic Plants (CSIR-CIMAP), Lucknow, Uttar Pradesh, India; Academy of Scientific and Innovative Research (AcSIR), Ghaziabad, Uttar Pradesh, India*
BOAZ NEGIN • *Institute of Plant Sciences and Genetics in Agriculture, The Hebrew University of Jerusalem, Rehovot, Israel; Department of Plant and Environmental Sciences, Alexander Silberman Institute of Life Sciences, The Hebrew University of Jerusalem, Jerusalem, Israel*
HANS-WILHELM NÜTZMANN • *Biosciences, University of Exeter, Exeter, UK*
ANNE OSBOURN • *Biochemistry and Metabolism Department, John Innes Centre, Norwich, UK; Department of Biochemistry and Metabolism, John Innes Centre, Norwich, UK*
KALLIOPE K. PAPADOPOULOU • *Laboratory of Plant and Environmental Biotechnology, Department of Biochemistry and Biotechnology, University of Thessaly, Larissa, Greece*

VIKTORIIA PLYHUN • *Department of Biology, National University of Kyiv-Mohyla Academy, Kyiv, Ukraine; Institute of Bioorganic Chemistry, Polish Academy of Sciences, Poznan, Poland*
GUY POLTURAK • *Institute of Plant Sciences and Genetics in Agriculture, The Hebrew University of Jerusalem, Rehovot, Israel*
KIERAN REYNOLDS • *Department of Life Sciences, University of Bath, Bath, UK*
AMIR RONEN • *Institute of Plant Sciences and Genetics in Agriculture, The Hebrew University of Jerusalem, Rehovot, Israel*
MICHAL ROSENTHAL • *Institute of Plant Sciences and Genetics in Agriculture, The Hebrew University of Jerusalem, Rehovot, Israel*
ANASTASIIA SATYR • *Institute of Bioorganic Chemistry, Polish Academy of Sciences, Poznan, Poland; Institute of Computing Science, Faculty of Computing and Telecommunications, Poznan University of Technology, Poznan, Poland*
ANJALI D. SHAH • *School of Biological Sciences, Life Sciences Building, University of Bristol, Bristol, UK; School of Chemistry, University of Bristol, Bristol, UK*
JASON SLOT • *Department of Plant Pathology, The Ohio State University, Columbus, OH, USA*
YUHUI SUN • *School of Pharmacy, Huazhong University of Science and Technology, Wuhan, China*
EVANGELOS C. TATSIS • *National Key Laboratory of Plant Molecular Genetics, CAS Center for Excellence in Molecular Plant Sciences, Shanghai Institute of Plant Physiology and Ecology, Chinese Academy of Sciences, Shanghai, China; CEPAMS – CAS-JIC Centre of Excellence for Plant and Microbial Sciences, Shanghai, China; School of Breeding and Multiplication (Sanya Institute of Breeding and Multiplication), Hainan University, Sanya, China*
JESSICA TAYLOR • *Department of Life Sciences, University of Bath, Bath, UK*
RAMESHA THIMMAPPA • *Amity Institute of Genome Engineering, Amity University, Noida, Uttar Pradesh, India*
PANAGIOTIS VLETSOS • *Laboratory of Plant and Environmental Biotechnology, Department of Biochemistry and Biotechnology, University of Thessaly, Larissa, Greece*
KATHERINE WILLIAMS • *School of Biological Sciences, Life Sciences Building, University of Bristol, Bristol, UK; School of Applied Sciences, University of the West of England, Bristol, UK*
QINAN WU • *College of Pharmacy, Nanjing University of Chinese Medicine, Nanjing, China; Jiangsu Collaborative Innovation Center of Chinese Medicinal Resources Industrialization, Nanjing University of Chinese Medicine, Nanjing, China*
SONG WU • *National Key Laboratory of Plant Molecular Genetics, CAS Center for Excellence in Molecular Plant Sciences, Shanghai Institute of Plant Physiology and Ecology, Chinese Academy of Sciences, Shanghai, China; University of Chinese Academy of Sciences, Shanghai, China*
ZHEYONG XUE • *State Key Laboratory of Rice Biology and Breeding, China National Center for Rice Improvement, China National Rice Research Institute, Hangzhou, China; Key Laboratory of Saline-alkali Vegetation Ecology Restoration, Ministry of Education, College of Life Science, Northeast Forestry University, Harbin, China*
XUE YIN • *Key Laboratory of Synthetic Biology of Ministry of Agriculture and Rural Affairs, Tobacco Research Institute, Chinese Academy of Agricultural Sciences, Qingdao, China*
ZHIHU ZHAO • *Department of Protein Engineering, Beijing Institute of Biotechnology, Beijing, China*

ZHIYU ZHONG • *School of Pharmacy, Huazhong University of Science and Technology, Wuhan, China; Key Laboratory of Combinatorial Biosynthesis and Drug Discovery (Ministry of Education), Wuhan University, Wuhan, China*
XUAN ZHOU • *Shanghai Collaborative Innovation Center of Agri-Seeds, School of Agriculture and Biology, Shanghai Jiao Tong University, Shanghai, China*
AGNIESZKA ZMIENKO • *Institute of Bioorganic Chemistry, Polish Academy of Sciences, Poznan, Poland; Institute of Computing Science, Faculty of Computing and Telecommunications, Poznan University of Technology, Poznan, Poland*

Chapter 1

Generalized Gene Cluster Detection Using *CLOCI*

Zachary Konkel and Jason Slot

Abstract

Metabolic gene clusters (MGCs) are genomic loci that contain multiple genes that are functionally and genetically linked. MGCs collectively encode a spectrum of metabolic functions, including small molecule biosynthesis, nutrient assimilation, metabolite degradation, and production of proteins essential for growth and development. Due to their diverse ecological functions, identifying gene clusters is a powerful tool for small molecule discovery and provides insight into the ecology and evolution of organisms. Gene cluster detection algorithms have historically been specialized for detecting biosynthetic gene clusters that contain canonical "core" biosynthetic functions, while overlooking uncommon or unknown cluster classes. These overlooked clusters are a potential source of novel natural products and comprise an untold portion of overall gene cluster repertoires. Unbiased, *function-agnostic* detection algorithms therefore provide an opportunity to reveal novel classes of gene clusters and more precisely define genome organization.

We developed *CLOCI* (Co-occurrence Locus and Orthologous Cluster Identifier) as a generalized, unbiased gene cluster detection algorithm. *CLOCI* generalizes gene cluster detection by identifying signatures of coordinated gene evolution that underlie all classes of MGCs. *CLOCI* first detects selection on gene colocalization by identifying and circumscribing shared synteny loci across a dataset of genomes into homologous locus groups. Gene clusters comprise a subset of these homologous locus groups, and *CLOCI* implements orthogonal proxies of coordinated gene evolution, such as quantifying loss and horizontal transfer of a locus, to enrich MGCs from homologous loci. Here, we describe the conceptual framework of the *CLOCI* algorithm and present a description of its implementation (*see* **Note 1**).

Key words Metabolic gene clusters, Biosynthetic, Detection, Synteny, Homologous locus, Fungi

1 Introduction

Metabolic gene clusters (MGCs) are chromosomal loci that contain two or more adjacent genes that are co-inherited and collectively encode a concerted metabolic phenotype. Gene clusters are found across the tree of life and encode diverse metabolic and ecological phenotypes [1–6]. Perhaps the most well-studied category of gene clusters is biosynthetic gene clusters, which produce small molecule "secondary/specialized metabolites" that mediate biotic and abiotic interactions. Specialized metabolites have diverse activities with applications in industry and agriculture [7] and are a prominent

Hans-Wilhelm Nützmann and Ramesha Thimmappa (eds.), *Metabolic Gene Clusters: Methods and Protocols*, Methods in Molecular Biology, vol. 3054, https://doi.org/10.1007/978-1-0716-5384-5_1,

source of new drugs in healthcare [8]. These compounds include UV-absorbing carotenoids [9], antimicrobial compounds [10, 11], and siderophores [12, 13]. While relatively overlooked, MGCs can also encode pathways that degrade antagonistic compounds [14], catabolize carbon sources and amino acids [15–17], assimilate nutrients [18], or produce essential vitamins [19] and proteins [20–22]. Due to their diverse functions, gene cluster detection has become a transformative tool for ecological genomics and specialized metabolite drug discovery in the genomic era.

To date, the most widely implemented gene cluster detection algorithms are function-centric [5, 23] and focus on biosynthetic MGCs. Function-centric algorithms search for profile models of "core" biosynthetic proteins commonly associated with described biosynthetic gene clusters of bacteria and filamentous Ascomycota (fungi). Profile models are largely constructed from a set of "canonical" core proteins derived from polyketide, nonribosomal peptide, post-translationally modified peptide, and terpene biosynthetic clusters [24]. These few models have been successful in uncovering a diversity of small-molecule biosynthetic pathways because canonical core functions produce polymer-like compounds that can incorporate diverse monomers, decorative moieties, and intramolecular bonding that yield diverse structures.

While function-centric detection may unveil small molecule biosynthetic clusters in lineages that prominently contain canonical MGCs, these algorithms have limited utility in lineages replete with "non-canonical" specialized metabolite gene clusters that lack core genes upon which the models rely [25–30]. Indeed, Greater than 94% of the fungal clusters in the Minimum Information about Biosynthetic Gene Clusters (MIBiG) database [24] originate from the filamentous fungi in the phylum Ascomycota, depicting bias in gene cluster identification. To account for non-canonical classes, the function-centric software *antiSMASH* incrementally expands its scope by accumulating models of novel biosynthetic gene cluster core functions [5]. Each update increases the breadth of future searches to include new biosynthetic classes, though the function-centric paradigm focuses on biosynthetic MGCs and cannot infer novel non-canonical gene cluster classes de novo, which precludes unbiased comparative genomic analysis of cluster repertoires. In order to unveil new small-molecule biosynthetic pathways and identify other classes of MGCs, a generalized, *function-agnostic* gene cluster detection algorithm is necessary.

All MGCs are formed and persist due to selection favoring coordinated gene evolution, and detecting universal signatures of shared selection can be used to generalize MGC detection. Gene clusters comprise cooperative genes that are tied to discrete phenotypes, and the clustered state is maintained by selection [31–33]. Gene clusters are evidence of selection because it is improbable that multiple distantly related genes convergently co-locate in diverse lineages and endure microsynteny decay over time [14, 31]. One proxy that identifies selection on gene order is unexpectedly shared microsynteny (synteny) [14], which detects selection for gene colocalization by identifying combinations of gene families that co-occur with one another in an unexpectedly large phylogenetic distribution. While detecting unexpectedly shared microsynteny can unveil gene clusters [34, 35], other types of genomic regions can meet unexpectedness thresholds because strong linkage disequilibrium within the cluster may extend to the surrounding regions [36]. Additionally, shared synteny loci that are not gene clusters, such as regions surrounding the centromere [37], may meet unexpectedness thresholds. However, MGCs experience manifestations of coordinated gene evolution that, when measured together, can distinguish them from other types of shared synteny loci:

- MGCs often have sparse phylogenetic distributions, indicative of horizontal transfer and loss. Entire MGCs may be rapidly lost when their encoded phenotype does not have a strong effect on fitness, and MGCs are amenable to horizontal transfer because they encode multigene phenotypes within discrete genomic regions [32, 38].
- Genes within MGCs have a high degree of commitment to the clustered locus. Once a gene becomes a part of an MGC, the gene's evolutionary trajectory is entangled with the overall MGC's with relatively rare exception, though recombination of MGCs can lead to genes moving between MGC families [29, 38, 39].
- MGCs are often under strong purifying selection due to the strong interactions their phenotypes have with the overall genome's fitness.

We created the Co-occurrence Locus and Orthologous Cluster Identifier (*CLOCI*) to generally detect MGCs using these measurements of coordinated gene evolution. While identifying selection on gene colocalization, *CLOCI* circumscribes groups of unexpectedly shared synteny loci by extending the logic underlying circumscribing homologous gene groups (orthogroups) to the locus level. *CLOCI* then filters MGCs from these homologous locus groups (HLGs) using proxies of selection on coordinated gene evolution. Briefly, the *CLOCI* algorithm operates by modeling the rate of microsynteny decay for inputted genomes; defining loci with shared microsynteny in extreme percentiles of the microsynteny decay null distributions; circumscribing homologous loci based on amino acid similarity and homologous gene overlap; and filtering MGCs from homologous loci using proxies of coordinated gene evolution, such as phylogenetic distribution sparsity and gene commitment to the locus. *CLOCI* yields the greatest recovery and categorical diversity of gene clusters when compared to *anti-SMASH* and *EvolClust* [40]. The *CLOCI* detection model has the additional advantage of accurate gene cluster boundary inference and predicts both novel putative clusters and non-canonical biosynthetic gene clusters [41].

2 Materials

CLOCI fundamentally requires multiple genomes to identify MGCs across the dataset, a microsynteny tree to accurately model microsynteny decay, and sufficient computing resources to run the algorithm. These genomes must be annotated and are ideally submitted using *Mycotools* [42], a genome database software that automates public genome acquisition and local genome data curation. The microsynteny tree depends on extracting representative gene neighborhoods that can model decay in gene order between genomes. A description of acquiring these data is noted in Methods below. Standard Unix-like computing workbenches are sufficient to run *CLOCI*, referencing several hundred genomes, though high-performance computing environments with access to large amounts of RAM are necessary for phylum-kingdom scale analyses. While most of the *CLOCI* algorithm is implemented using matrix-based operations, the overall compute time is superlinear relative to input size because some steps involve pairwise comparisons. The RAM requirements for *CLOCI* will vary drastically depending on the number of inputted genomes, the size of these genomes, and the initial thresholds that affect the quantity of analyzed loci. *CLOCI* has run successfully for 100s of genomes using 64 GB of RAM on a local workbench, while also successfully completing with 1000s of genomes using 1 TB of RAM in a high-performance computing environment.

3 Methods

3.1 Overview

The workflow for *CLOCI* is illustrated in Fig. 1 and described here: (1) Generate a MycotoolsDB with genomes of interest [42]; (2) generate a robust microsynteny tree referencing single-copy orthologs gene neighborhoods and constraining the topology using a reference phylogenomic tree to constrain the topology of a microsynteny tree; (3) run *CLOCI*; and (4) add filtering parameters to *CLOCI* output to extract gene cluster families (GCFs) from homologous locus groups (HLGs).

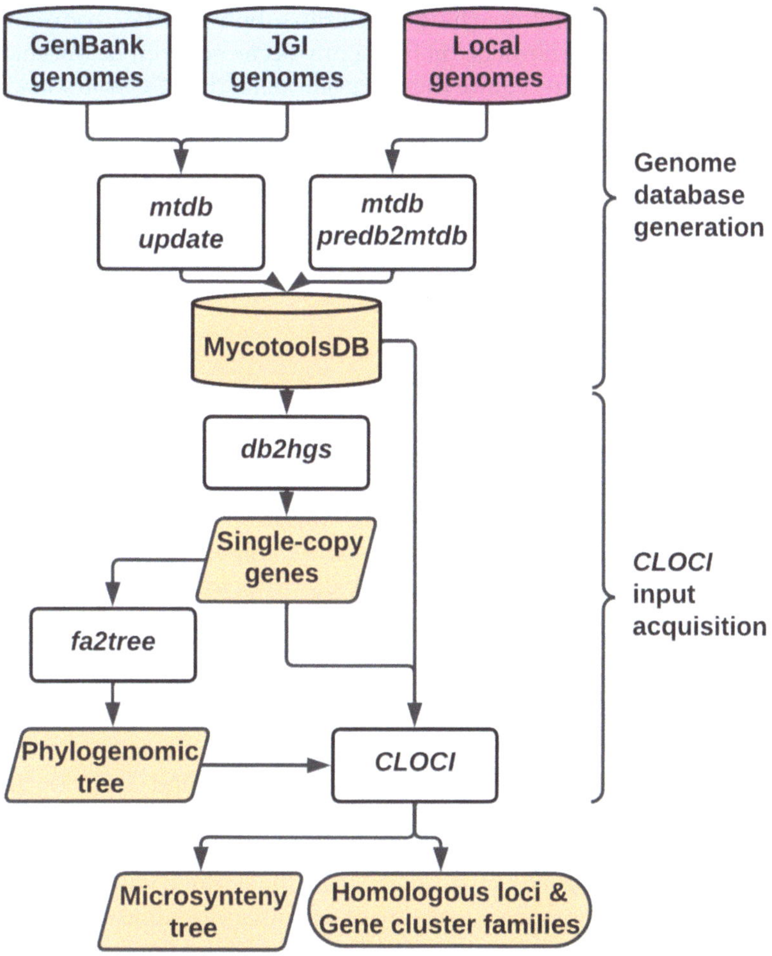

Fig. 1 Flowchart depicting *CLOCI* analysis workflow. Users acquire genomes for the analysis using *Mycotools* to acquire publicly available and local genome data. Following database constructions, users use *Mycotools* scripts to build a phylogenomic tree and acquire single-copy gene homology groups for *CLOCI* input

3.2 Acquiring an Input Genome Dataset

CLOCI requires a dataset of multiple input genomes to recover MGCs. It is important to adequately sample a lineage's genomes because MGCs will only be recovered if their phylogenetic distribution of conserved microsynteny lies within the upper percentile of null distributions of microsynteny decay. It is generally recommended to run *CLOCI* at a taxonomic level that will recover horizontal transfer events because horizontal transfer drastically increases the phylogenetic distance of conserved microsynteny. The distance with which horizontal transfer occurs decreases relative to phylogenetic distance between genomes, but the distance of common transfer events varies depending on the lineage studied. For example, in the phylum Basidiomycota (fungi), sampling a taxonomic order is likely sufficient [29, 43], while in the phylum

Ascomycota (fungi), and the subphylum Pezizomycotina in particular, transfers can frequently occur between taxonomic classes [39].

To acquire these genome data, the *Mycotools* software suite [42] enables scalable, automated acquisition of lineage-wide genome data and is directly integrated with *CLOCI*. A full description of acquiring lineages and integrating local genome data is available at the *Mycotools* repository wiki, though a current example command for acquiring publicly available genome data from a fungal taxonomic class would be:

```
mtdb update --init <INITIALIZATION_DIR> --lineage Agaricomy-
cetes --rank class
```

This command will create a *Mycotools* database directory with genome FASTAs, gene annotation GFFs, and a relational tab-delimited database, ".mtdb" file, which relates accessions on each row to their respective genomic data files and metadata. The ".mtdb" file is used as the downstream input to *CLOCI*.

3.3 Generating a Robust Microsynteny Tree

To accurately model microsynteny decay, *CLOCI* generates a microsynteny tree that recapitulates the divergence in shared gene order between genomes. It is recommended to initialize *CLOCI* referencing a dataset of single-copy ortholog gene neighborhoods to model microsynteny decay from and a phylogenomic tree to constrain the topology of the microsynteny tree. Single-copy orthologs are ideal for microsynteny tree reconstruction because the genes are present across the entire dataset, the gene neighborhoods they reside in are presumed to be relatively stable, and they are not influenced by selection acting on duplicates. Additionally, *CLOCI* will also receive an inputted phylogenomic tree generated from these single-copy orthologs because a phylogenomic tree provides a robust constraint to the topology of the microsynteny tree, which ensures that the relationships between genomes in the microsynteny tree are accurately depicted.

CLOCI attempts to automatically detect near single-copy gene family neighborhoods for reconstructing the microsynteny tree, though it is recommended to explicitly input these near single-copy genes using the -f argument referencing a file of reference genes. Single-copy genes are tractable markers for microsynteny tree reconstruction because they are found in genomes across the dataset and they are frequently duplicated, which would confound identifying orthologous gene neighborhoods. RNA polymerase-II subunit B, or *RPB2*, is a prominent single-copy gene found across most fungi. Single-copy genes can be inferred by *OrthoFinder* [44], near single-copy genes determined by *db2hgs* [42], or searching for a precompiled set of marker genes that are in near single-copy abundance across the inputted genomes. *OrthoFinder* can become intractable with large datasets, so two examples of *db2hgs* usage are presented in the following commands, which will output a directory of single-copy or near single-copy gene FASTAs using the genomes depicted in the ".mtdb" database generated previously:

```
# Identify single-copy genes
db2hgs -d <MTDB_file>

# Identify near-single copy genes using standard deviation,
median, and percent of genomes
db2hgs -d <MTDB_file> --stdev <MAX_STDEV> --median <MAX_ME-
DIAN> --min_genomes <MIN_PERCENT_OF_GENOMES>
```

Single-copy genes are also tractable models for a multi-gene phylogenomic tree, which is an ideal input to constrain the microsynteny tree topology. Single-copy genes are ideally suited for phylogenomic tree reconstruction because these genes are expected to be congruent with the evolution of the genome and not commonly affected by duplication and horizontal transfer. A full description of phylogenomic tree reconstruction is beyond the scope of this method description, though *Mycotools* incorporates a set of modules that simplify this process and are outlined in its usage wiki.

3.4 Selecting Filtering Parameters for Extracting Gene Clusters from Homologous Loci

As of the date of this publication, *CLOCI* default parameters have been tuned for our initial dataset on ~2247 fungi across the kingdom [41]. These should suffice for circumscribing homologous loci in most analyses, though the gene cluster family filtering parameters are ideally determined referencing known clusters from your particular dataset. By default, thresholds for all proxies of coordinated gene evolution are set to 0 because these thresholds will vary for the type of clusters of interest and the lineage, though some common reference cluster values and their impact on filtering diverse MGCs can be visualized in Fig. 2 (*see* **Note 2**). We recommend compiling a dataset of known cluster reference genes, running *CLOCI*, identifying those genes in the output, determining the values for the reference cluster proxies, and then implementing the thresholds. We suspect our pilot study reached a local maximum in terms of output quality, though a global maximum may be achieved with future machine learning-based parameter tuning.

Gene cluster filtering parameters are based on proxies for selection on coordinated gene evolution and involve math and explanations elaborated on in the methods and supplementary section of the original *CLOCI* manuscript [41]. Below is an example of running *CLOCI* with minimally supplied inputs, followed by subsequent gene cluster extraction by filtering with proxies of coordinated gene evolution:

```
# Extract single-copy gene accessions from OrthoFinder/db2hgs'
single-copy FASTA directory
```

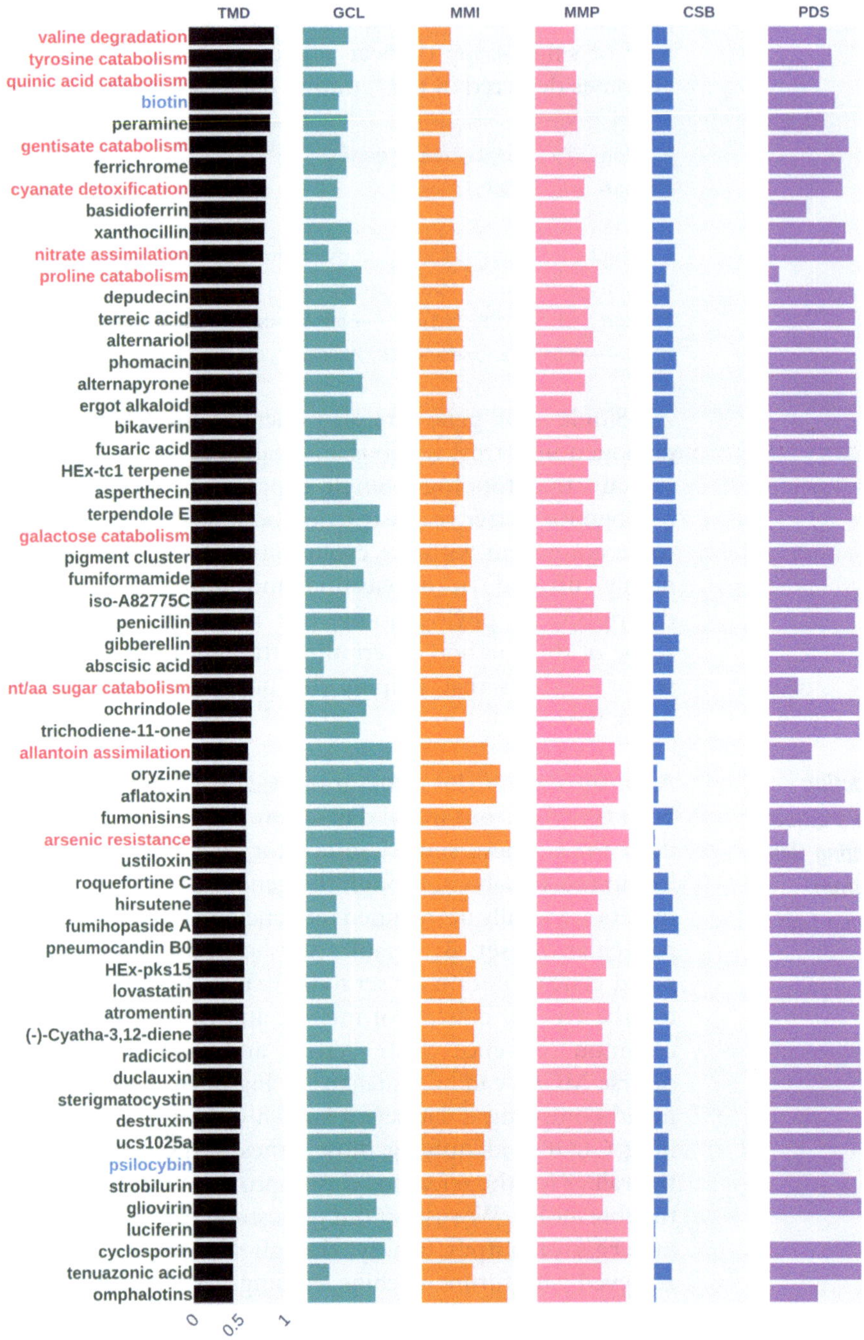

Fig. 2 Tuning parameters for target functions using proxies of coordinated gene evolution for reference gene clusters. Canonical biosynthetic clusters' text is highlighted in purple, non-canonical biosynthetic clusters are highlighted blue, and non-biosynthetic clusters are highlighted in red. Values of proxies of coordinated gene evolution are percentages or normalized for total microsynteny distance (TMD), gene commitment to the locus (GCL), mean minimum percent identity (MMI), mean minimum percent positives (MMP), conservative substitution bias (CSB), and phylogenetic distribution sparsity (PDS)

```
grep -h ">" <SINGLE_COPY_FASTA_DIR>/* | grep -Po "[^>]+" >
sco_accessions.txt

# Run CLOCI on a full preassembled MycotoolsDB with a micro-
synteny tree constrained by a reference phylogenomic tree:
cloci --root "<GENOME_1>,<GENOME_2>" \
 -c <PHYLOGENOMIC_TREE> \
 -f sco_accessions.txt

# Once complete, evaluate and extract GCFs using proxies of
coordinated gene evolution:
cloci -o <PREVIOUS_CLOCI_OUTPUT_DIR> \
 -pt <PDS_THRESHOLD> -ct <CSB_THRESHOLD> \
 -gt <GCL_THRESHOLD> -tt <TMD_THRESHOLD> \
 --cpus <CPUS> --root "<GENOME_1>,<GENOME_2>" \
 -c <PHYLOGENOMIC_TREE_CONSTRAINT> \
 -f sco_accessions.txt
```

Below is a brief explanation on the proxies of coordinated gene evolution; please refer to the *CLOCI* manuscript for further elaboration:

--id_percent: Mean minimum percent identity (MMI) of shared gene families within a gene cluster family (GCF).

--pos_percent: Mean minimum percent positive (MMP) alignment substitutions of shared gene families within a GCF.

--csb_threshold: Minimum conservative substitution bias for GCFs. Conservative substitution bias is a proxy for detecting purifying selection as the ratio of structurally conservative (positive) amino acid substitutions to identical amino acid substitutions.

--pds_threshold: GCF phylogenetic distribution sparsity is a measurement of the percent of the branch length descended from the most recent shared ancestor of the genomes with the GCF that contains the GCF.

--gcl_treshold: GCF gene commitment to the locus is an alignment-based quantification of the degree to which a gene family is committed to the clustered locus. Fully committed gene families have the most similar alignments with all other representatives of the gene family in the GCF.

4 Notes

1. We describe *CLOCI* as an extension of homologous (orthologous) gene grouping to the locus level [41, 44], which inherently comprises loci that are not gene clusters. This means *CLOCI* serves a broader role than MGC identification in

describing genome organization, and it also indicates that the output will contain loci that may not be MGCs. To this end, the orthogonal filtering proxies of coordinated gene evolution can selectively enrich MGCs from *CLOCI* HLGs (Fig. 2), though functional validation is always necessary to corroborate MGC function.

CLOCI currently largely generates text-based output that depicts homologous locus groups (HLGs) predicted for the whole dataset, and genome-level HLG predictions. In addition to HLG files, equivalent gene cluster family (GCF) files are outputted if GCF filtering parameters are used. The whole dataset HLGs file, "hlgs.tsv.gz," contains a line-by-line delineation of HLGs, the values for each proxy of coordinated gene evolution, and the genomes the HLG is recovered in. The genome-specific HLG files contain a line-by-line delineation of loci recovered in the genome, including their HLG identification number, genes within the HLG, and abbreviated Pfam/InterPro [45, 46] functional annotations of the recovered genes. Multiple loci can be circumscribed into a single HLG, such as the case with contig-edge clusters.

Extracting anchor/core genes and other functions of interest is an alternative method of MGC discovery that couples function-centric detection with *CLOCI*'s shared synteny support and increased cluster boundary accuracy. *CLOCI* includes an auxiliary tool, *cloci2annotate*, that can extract HLGs that contain interesting functions. For example, this approach can use *antiSMASH*'s secondary metabolite core gene hidden Markov models to incorporate the benefits of *CLOCI*'s increased cluster boundary accuracy and orthogonal support for cluster calling from shared synteny locus detection. Alternatively, users can extract a list of functions that correspond to predicted MGC metabolic pathways, which enables the discovery of non-canonical MGCs.

2. *CLOCI* comes packaged with several tools that facilitate extracting and interpreting data for downstream analysis. Each script is executed following a *CLOCI* run and will output a subdirectory within the reference *CLOCI* output directory.

   cloci2stats

   cloci2stats provides a taxonomy-binned summary of output statistics for HLG/GCF predictions. The output includes files summarizing the mean and median values for proxies of coordinated gene evolution, genes per cluster, percent of genome clustered, and basic assembly statistics for inputted genomes.

   hlg2biofile

   hlg2biofile outputs biological data file formats, such as GenBanks, FastAs, and GFFs, for all representative loci of specific genomes, HLGs, or GCFs.

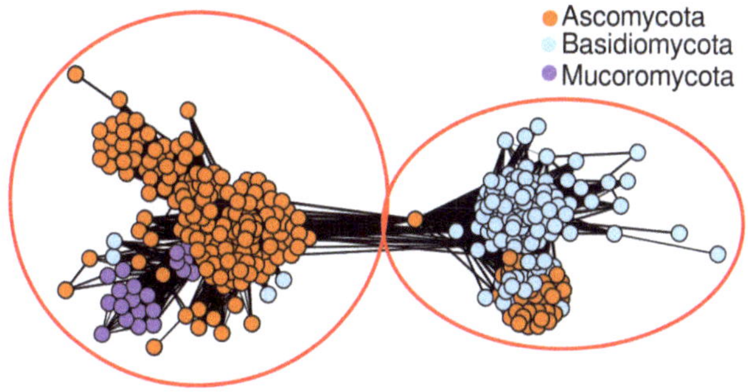

Fig. 3 Network of two nitrate assimilation gene cluster families (GCFs) generated via *hlg2hlg_net*. Nodes depict a locus, node colors depict phyla, edge weight/distance depicts locus-locus similarity, and red circles highlight the two different GCFs. The proximity of variously colored nodes indicates highly similar clusters are detected in distantly related taxa, which suggests that horizontal transfer is contributing to the distribution of these gene clusters. For example, Ascomycota nodes are found nested within distinct portions of this graph, which suggests that some Ascomycota nitrate assimilation clusters may have independently originated from horizontal transfer instead of vertical inheritance. Several Mucoromycota nodes are found nested near Basidiomycota and Ascomycota, which also present potential for horizontal transfer to this phylum. This graph is based on locus-locus similarity, which is insufficient on its own to conclusively determine horizontal transfer, though phylogenetic analysis of the genes within these clusters has indeed corroborated horizontal transfer [41, 47]

cloci2annotate

cloci2annotate will identify and extract HLGs/GCFs that contain a hit to an inputted set of hidden Markov models (HMMs).

hlg2hlg_net

hlg2hlg_net generates visualizations of locus-locus similarity networks for specified HLGs/GCFs. The edge weights of this network depict the similarity between loci used in the initial HLG circumscription. Optionally, node colors can depict taxonomic lineage or HLG ID. An example of a network depicting two interconnected GCFs is shown in Fig. 3.

References

1. Mihali TK, Kellmann R, Neilan BA (2009) Characterisation of the paralytic shellfish toxin biosynthesis gene clusters in Anabaena circinalis AWQC131C and Aphanizomenon sp. NH-5. BMC Biochem 10:8. https://doi.org/10.1186/1471-2091-10-8
2. Burger G, Gray MW, Forget L, Lang BF (2013) Strikingly bacteria-like and gene-rich mitochondrial genomes throughout jakobid protists. Genome Biol Evol 5:418–438. https://doi.org/10.1093/gbe/evt008
3. Nützmann H-W, Osbourn A (2014) Gene clustering in plant specialized metabolism. Curr Opin Biotechnol 26:91–99. https://doi.org/10.1016/j.copbio.2013.10.009
4. Keller NP (2015) Translating biosynthetic gene clusters into fungal armor and weaponry. Nat Chem Biol 11:671–677. https://doi.org/10.1038/nchembio.1897

5. Blin K, Shaw S, Kloosterman AM, Charlop-Powers Z, van Wezel GP, Medema MH, Weber T (2021) antiSMASH 6.0: improving cluster detection and comparison capabilities. Nucleic Acids Res 49:W29–W35. https://doi.org/10.1093/nar/gkab335
6. Ettema TJG, Brinkman AB, Lamers PP, Kornet NG, de Vos WM, van der Oost J (2006) Molecular characterization of a conserved archaeal copper resistance (cop) gene cluster and its copper-responsive regulator in Sulfolobus solfataricus P2. Microbiology 152:1969–1979. https://doi.org/10.1099/mic.0.28724-0
7. Nofiani R, de Mattos-Shipley K, Lebe KE, Han L-C, Iqbal Z, Bailey AM, Willis CL, Simpson TJ, Cox RJ (2018) Strobilurin biosynthesis in Basidiomycete fungi. Nat Commun 9:1–11. https://doi.org/10.1038/s41467-018-06202-4
8. Newman DJ, Cragg GM (2020) Natural products as sources of new drugs over the nearly four decades from 01/1981 to 09/2019. J Nat Prod 83:770–803. https://doi.org/10.1021/acs.jnatprod.9b01285
9. Linnemannstöns P, Prado M, Fernández-Martín R, Tudzynski B, Avalos J (2002) A carotenoid biosynthesis gene cluster in Fusarium fujikuroi: the genes carB and carRA. Mol Gen Genomics 267:593–602. https://doi.org/10.1007/s00438-002-0690-5
10. Alberti F, Khairudin K, Venegas ER, Davies JA, Hayes PM, Willis CL, Bailey AM, Foster GD (2017) Heterologous expression reveals the biosynthesis of the antibiotic pleuromutilin and generates bioactive semi-synthetic derivatives. Nat Commun 8:1831. https://doi.org/10.1038/s41467-017-01659-1
11. Díez B, Gutiérrez S, Barredo JL, van Solingen P, van der Voort LH, Martín JF (1990) The cluster of penicillin biosynthetic genes. Identification and characterization of the pcbAB gene encoding the alpha-aminoadipyl-cysteinyl-valine synthetase and linkage to the pcbC and penDE genes. J Biol Chem 265:16358–16365
12. Brandenburger E, Gressler M, Leonhardt R, Lackner G, Habel A, Hertweck C, Brock M, Hoffmeister D (2017) A highly conserved basidiomycete peptide synthetase produces a trimeric hydroxamate siderophore. Appl Environ Microbiol 83:e01478–e01417. https://doi.org/10.1128/AEM.01478-17
13. Perrin RM, Fedorova ND, Bok JW, Cramer RA Jr, Wortman JR, Kim HS, Nierman WC, Keller NP (2007) Transcriptional regulation of chemical diversity in Aspergillus fumigatus by LaeA. PLoS Pathog 3:e50. https://doi.org/10.1371/journal.ppat.0030050
14. Gluck-Thaler E, Slot JC (2018) Specialized plant biochemistry drives gene clustering in fungi. ISME J 12:1694–1705. https://doi.org/10.1038/s41396-018-0075-3
15. Arst HN, MacDonald DW (1975) A gene cluster in Aspergillus nidulans with an internally located cis-acting regulatory region. Nature 254:26–31. https://doi.org/10.1038/254026a0
16. Douglas HC, Hawthorne DC (1966) Regulation of genes controlling synthesis of the galactose pathway enzymes in yeast. Genetics 54: 911–916. https://doi.org/10.1093/genetics/54.3.911
17. Greene GH, McGary KL, Rokas A, Slot JC (2014) Ecology drives the distribution of specialized tyrosine metabolism modules in fungi. Genome Biol Evol 6:121–132. https://doi.org/10.1093/gbe/evt208
18. Johnstone IL, McCabe PC, Greaves P, Gurr SJ, Cole GE, Brow MAD, Unkles SE, Clutterbuck AJ, Kinghorn JR, Innis MA (1990) Isolation and characterisation of the crnA-niiA-niaD gene cluster for nitrate assimilation in Aspergillus nidulans. Gene 90:181–192. https://doi.org/10.1016/0378-1119(90)90178-T
19. Magliano P, Flipphi M, Sanglard D, Poirier Y (2011) Characterization of the Aspergillus nidulans biotin biosynthetic gene cluster and use of the bioDA gene as a new transformation marker. Fungal Genet Biol 48:208–215. https://doi.org/10.1016/j.fgb.2010.08.004
20. Fritsch EF, Lawn RM, Maniatis T (1980) Molecular cloning and characterization of the human β-like globin gene cluster. Cell 19:959–972. https://doi.org/10.1016/0092-8674(80)90087-2
21. Forrester WC, Thompson C, Elder JT, Groudine M (1986) A developmentally stable chromatin structure in the human beta-globin gene cluster. Proc Natl Acad Sci 83:1359–1363. https://doi.org/10.1073/pnas.83.5.1359
22. Lewis EB (1978) A gene complex controlling segmentation in Drosophila. Nature 276:565–570. https://doi.org/10.1038/276565a0
23. Yaegashi J, Oakley BR, Wang CCC (2014) Recent advances in genome mining of secondary metabolite biosynthetic gene clusters and the development of heterologous expression systems in Aspergillus nidulans. J Ind Microbiol Biotechnol 41:433–442. https://doi.org/10.1007/s10295-013-1386-z
24. Terlouw BR, Blin K, Navarro-Muñoz JC, Avalon NE, Chevrette MG, Egbert S, Lee S, Meijer D, Recchia MJJ, Reitz ZL, van Santen

JA, Selem-Mojica N, Tørring T, Zaroubi L, Alanjary M, Aleti G, Aguilar C, Al-Salihi SAA, Augustijn HE, Avelar-Rivas JA, Avitia-Domínguez LA, Barona-Gómez F, Bernaldo-Agüero J, Bielinski VA, Biermann F, Booth TJ, Carrion Bravo VJ, Castelo-Branco R, Chagas FO, Cruz-Morales P, Du C, Duncan KR, Gavriilidou A, Gayrard D, Gutiérrez-García K, Haslinger K, Helfrich EJN, van der Hooft JJJ, Jati AP, Kalkreuter E, Kalyvas N, Kang KB, Kautsar S, Kim W, Kunjapur AM, Li Y-X, Lin G-M, Loureiro C, Louwen JJR, Louwen NLL, Lund G, Parra J, Philmus B, Pourmohsenin B, Pronk LJU, Rego A, Balaya Rex DA, Robinson S, Rosas-Becerra LR, Roxborough ET, Schorn MA, Scobie DJ, Singh KS, Sokolova N, Tang X, Udwary D, Vigneshwari A, Vind K, Vromans SPJM, Waschulin V, Williams SE, Winter JM, Witte TE, Xie H, Yang D, Yu J, Zdouc M, Zhong Z, Collemare J, Linington RG, Weber T, Medema MH (2022) MIBiG 3.0: a community-driven effort to annotate experimentally validated biosynthetic gene clusters. Nucleic Acids Res. https://doi.org/10.1093/nar/gkac1049

25. Hewage RT, Aree T, Mahidol C, Ruchirawat S, Kittakoop P (2014) One strain-many compounds (OSMAC) method for production of polyketides, azaphilones, and an isochromanone using the endophytic fungus Dothideomycete sp. Phytochemistry 108:87–94. https://doi.org/10.1016/j.phytochem.2014.09.013
26. Medema MH, Blin K, Cimermancic P, de Jager V, Zakrzewski P, Fischbach MA, Weber T, Takano E, Breitling R (2011) antiSMASH: rapid identification, annotation and analysis of secondary metabolite biosynthesis gene clusters in bacterial and fungal genome sequences. Nucleic Acids Res 39:W339–W346. https://doi.org/10.1093/nar/gkr466
27. Fricke J, Blei F, Hoffmeister D (2017) Enzymatic synthesis of psilocybin. Angew Chem Int Ed 56:12352–12355. https://doi.org/10.1002/anie.201705489
28. Obermaier S, Müller M (2020) Ibotenic acid biosynthesis in the fly agaric is initiated by glutamate hydroxylation. Angew Chem Int Ed 59:12432–12435. https://doi.org/10.1002/anie.202001870
29. Reynolds HT, Vijayakumar V, Gluck-Thaler E, Korotkin HB, Matheny PB, Slot JC (2018) Horizontal gene cluster transfer increased hallucinogenic mushroom diversity. Evol Lett 2:88–101. https://doi.org/10.1002/evl3.42
30. Voigt K, Wolf T, Ochsenreiter K, Nagy G, Kaerger K, Shelest E, Papp T (2016) 15 Genetic and metabolic aspects of primary and secondary metabolism of the zygomycetes. In: Hoffmeister D (ed) Biochemistry and molecular biology. Springer International Publishing, Cham, pp 361–385
31. Louwen JJR, Kautsar SA, van der Burg S, Medema MH, van der Hooft JJJ (2023) iPRESTO: automated discovery of biosynthetic sub-clusters linked to specific natural product substructures. PLoS Comput Biol 19:e1010462. https://doi.org/10.1371/journal.pcbi.1010462
32. Lawrence JG, Roth JR (1996) Selfish operons: horizontal transfer may drive the evolution of gene clusters. Genetics 143:1843–1860
33. McGary KL, Slot JC, Rokas A (2013) Physical linkage of metabolic genes in fungi is an adaptation against the accumulation of toxic intermediate compounds. PNAS 110:11481–11486. https://doi.org/10.1073/pnas.1304461110
34. Gluck-Thaler E, Haridas S, Binder M, Grigoriev IV, Crous PW, Spatafora JW, Bushley K, Slot JC (2020) The architecture of metabolism maximizes biosynthetic diversity in the largest class of fungi. Mol Biol Evol 37:2838–2856. https://doi.org/10.1093/molbev/msaa122
35. Winter S, Jahn K, Wehner S, Kuchenbecker L, Marz M, Stoye J, Böcker S (2016) Finding approximate gene clusters with Gecko 3. Nucleic Acids Res 44:9600–9610. https://doi.org/10.1093/nar/gkw843
36. Rose CJ, Chapman JR, Marshall SDG, Lee SF, Batterham P, Ross HA, Newcomb RD (2011) Selective sweeps at the organophosphorus insecticide resistance locus, Rop-1, have affected variation across and beyond the α-esterase gene cluster in the Australian sheep blowfly, Lucilia cuprina. Mol Biol Evol 28:1835–1846. https://doi.org/10.1093/molbev/msr006
37. Douglass AP, Byrne KP, Wolfe KH (2019) The Methylotroph Gene Order Browser (MGOB) reveals conserved synteny and ancestral centromere locations in the yeast family Pichiaceae. FEMS Yeast Res 19:foz058. https://doi.org/10.1093/femsyr/foz058
38. Rokas A, Wisecaver JH, Lind AL (2018) The birth, evolution and death of metabolic gene clusters in fungi. Nat Rev Microbiol 16:731–744. https://doi.org/10.1038/s41579-018-0075-3
39. Scott K, Konkel Z, Gluck-Thaler E, David GEV, Simmt CF, Grootmyers D, Chaverri P, Slot J (2023) Endophyte genomes support greater metabolic gene cluster diversity compared with non-endophytes in Trichoderma. bioRxiv:2023.03.14.532605

40. Marcet-Houben M, Gabaldón T (2020) EvolClust: automated inference of evolutionary conserved gene clusters in eukaryotes. Bioinformatics 36(4):1265–1266. https://doi.org/10.1093/bioinformatics/btz706. PMID: 31560365; PMCID: PMC7703780
41. Konkel Z, Kubatko L, Slot JC (2024) CLOCI: unveiling cryptic fungal gene clusters with generalized detection. Nucleic Acids Res 52: e75. https://doi.org/10.1093/nar/gkae625
42. Konkel Z, Slot JC (2023) Mycotools: an automated and scalable platform for comparative genomics. bioRxiv:2023.09.08.556886
43. Luo H, Cai Q, Lüli Y, Li X, Sinha R, Hallen-Adams HE, Yang ZL (2018) The MSDIN family in amanitin-producing mushrooms and evolution of the prolyl oligopeptidase genes. IMA Fungus 9:225–242. https://doi.org/10.5598/imafungus.2018.09.02.01
44. Emms DM, Kelly S (2019) OrthoFinder: phylogenetic orthology inference for comparative genomics. Genome Biol 20:238. https://doi.org/10.1186/s13059-019-1832-y
45. El-Gebali S, Mistry J, Bateman A, Eddy SR, Luciani A, Potter SC, Qureshi M, Richardson LJ, Salazar GA, Smart A, Sonnhammer ELL, Hirsh L, Paladin L, Piovesan D, Tosatto SCE, Finn RD (2019) The Pfam protein families database in 2019. Nucleic Acids Res 47: D427–D432. https://doi.org/10.1093/nar/gky995
46. Jones P, Binns D, Chang H-Y, Fraser M, Li W, McAnulla C, McWilliam H, Maslen J, Mitchell A, Nuka G, Pesseat S, Quinn AF, Sangrador-Vegas A, Scheremetjew M, Yong S-Y, Lopez R, Hunter S (2014) InterProScan 5: genome-scale protein function classification. Bioinformatics 30:1236–1240. https://doi.org/10.1093/bioinformatics/btu031
47. Slot JC, Hibbett DS (2007) Horizontal transfer of a nitrate assimilation gene cluster and ecological transitions in fungi: a phylogenetic study. PLoS One 2(10). https://doi.org/10.1371/journal.pone.0001097

Chapter 2

In Silico Identification and Comparative Synteny of Biosynthetic Gene Clusters in Plants

Song Wu and Evangelos C. Tatsis

Abstract

Biosynthetic gene clusters (BGCs) are often encoding specialized metabolic pathways in plants, yet effective methods for the comparison across multiple species are still evolving. In this protocol, we present a bioinformatics approach combining plantiSMASH and MCScan for the identification, annotation, and comparative analysis of BGCs in plant genomes. The methodology involves using plantiSMASH to predict and annotate potential BGCs, followed by the use of MCScan to perform syntenic analysis, enabling the exploration of the conservation and evolutionary dynamics of BGCs across different plant species. Here, we provide a step-by-step guide for installing and configuring the necessary software, preparing genomic data, and executing the analysis. This methodology, integrated with transcriptomic and metabolomic data, can be used to verify the functional relevance of the identified BGCs in specific biosynthetic pathways. It is applicable to a broad range of plant species and serves as a framework for the discovery and characterization of BGCs. The described methodology can significantly enhance research in plant genomics and metabolic engineering by offering new insights into the organization and function of BGCs.

Key words Plant genomes, Biosynthetic gene clusters (BGCs), Comparative genomics, Syntenic analysis, Specialized metabolism, plantiSMASH, MCScan

1 Introduction

Biosynthetic gene clusters (BGCs) are pivotal for the synthesis of specialized metabolites in plants [1–3], contributing to their ecological adaptation, defense mechanisms, and medicinal properties [4]. The clustering of genes involved in biosynthetic pathways enables the coordinated expression and efficient synthesis of complex natural products. Biosynthetic gene clusters in plants can serve as "metabolic toolbox" providing, through gene duplications and mutations, new enzymes for metabolic diversification like the example of clerodane biosynthesis [5]. The clustering of metabolic genes provides an opportunity to map the evolutionary origins of the specific metabolic pathway in a more convenient way than for single genes scattered around the plant genome [6, 7]. Despite

Hans-Wilhelm Nützmann and Ramesha Thimmappa (eds.), *Metabolic Gene Clusters: Methods and Protocols*, Methods in Molecular Biology, vol. 3054, https://doi.org/10.1007/978-1-0716-5384-5_2,

progress in identifying BGCs in plant genomes, the evolutionary dynamics and regulatory mechanisms driving their formation and diversification are not fully understood.

Recent advancements in comparative genomics and synteny analysis have provided valuable insights into the conservation and diversification of BGCs across different plant lineages [6, 8]. However, systematic methods for the identification and comparison of these clusters, particularly in medicinal plants, are still evolving. Medicinal plants often contain unique BGCs that are responsible for the biosynthesis of bioactive compounds with significant pharmacological potential [9, 10]. Understanding the organization and evolutionary history of these clusters is crucial for elucidating the molecular basis of specialized metabolism.

To address these challenges, a cohesive bioinformatics approach is presented here (Fig. 1), combining plantiSMASH [11, 12] for BGC identification and MCScan [13, 14] for syntenic analysis across multiple plant genomes. This method enables efficient detection, annotation, and comparative analysis of BGCs, providing a comprehensive framework for exploring their conservation and divergence. By integrating further transcriptomic and

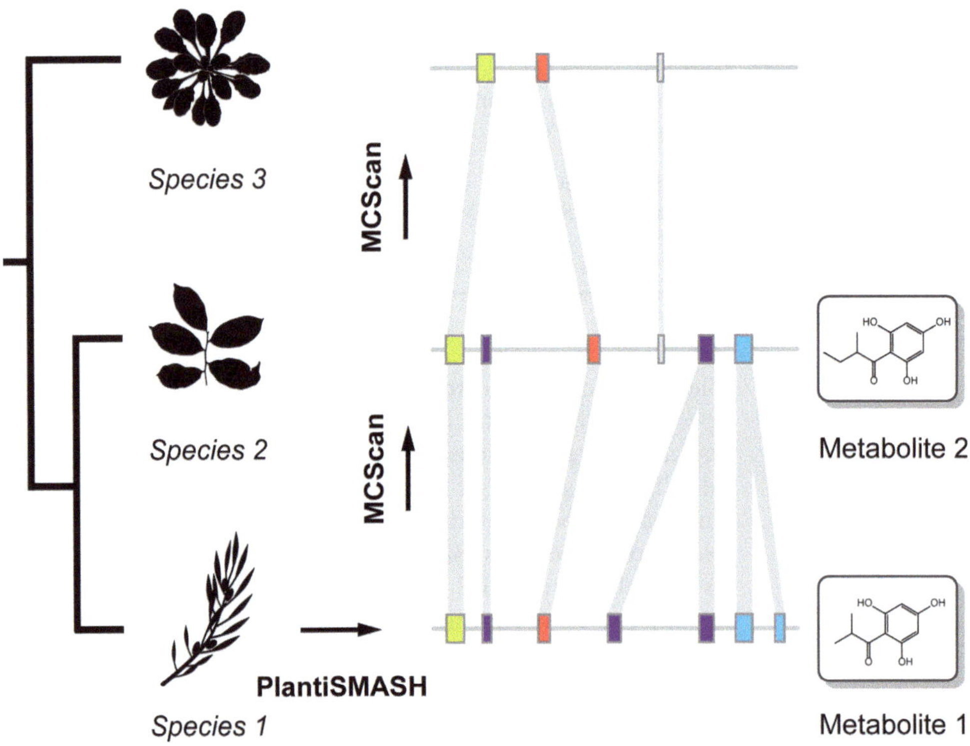

Fig. 1 A schematic representation of the applied methodology, integrating the detection of biosynthetic gene clusters with syntenic analysis, guided by further phylogenetic and phytochemical information, to decode the biosynthetic pathways of natural products

metabolomic data, this approach facilitates the functional validation of BGCs in specific biosynthetic pathways, linking gene organization with metabolite biosynthesis [6].

The value of this method is demonstrated through its application to several medicinal plant species. For example, the hyperforin BGC in *Hypericum perforatum* reveals how gene duplication and cluster compaction contribute to the biosynthesis of this pharmacologically active compound [6]. In Lamiaceae species, BGCs associated with ferruginol biosynthesis demonstrate how specific genomic rearrangements drive the production of bioactive diterpenes while tandem duplications provide nonfunctionalized enzymes [5]. These case studies illustrate the versatility of this approach in addressing complex questions related to BGC evolution and function.

This protocol provides a practical framework for the systematic exploration of BGCs across diverse plant species. It can be readily adapted to investigate the genetic basis of specialized metabolism in other medicinal plants and to identify new candidate BGCs for further functional and biochemical studies.

2 Software

1. Conda, a package and environment management system, version 23.1.0, or more recent.
2. plantiSMASH [11], software for identification, annotation, and analysis of biosynthesis gene clusters tailored to plant genomes, version 1.0.
3. JCVI [14], toolkit for comparative genomics analysis, version 1.0.6, or more recent.
4. Kaks_Calculator [15], software for the calculation of nonsynonymous and synonymous substitution rates between genes, version 2.0.

3 Methods

3.1 Installing and Utilization of Standalone Version plantiSMASH

plantiSMASH is a user-friendly tool for identifying potential biosynthetic gene clusters involved in specialized metabolism. It offers two usage options: an online platform (available at plantismash.secondarymetabolites.org) and a standalone version. While the web server provides convenience, the standalone version is preferred (*see* **Notes 1**, **2**, and **3**) for large-scale or complex analyses due to factors such as server load, downstream analysis requirements, and input file format considerations. Proper file formatting is crucial (*see* **Notes 4** and **5**), as incorrect formats can trigger errors or warnings that disrupt command execution—issues that are particularly problematic on external servers. Additionally, for large genomes with numerous coding genes, the standalone version can be resource-intensive, often consuming significant memory and leading to frequent errors. This makes local installation and careful resource management essential when handling large datasets.

1. Create a dedicated Conda environment to ensure compatibility of plantiSMASH.

```
conda create -n plantismash
conda activate plantismash
```

2. Download the plantiSMASH package from the official repository.

```
git clone https://bitbucket.org/satriaphd/plantismash-re-
lease.git
```

3. Prepare a requirements.txt file that includes all the dependencies.

```
python=2.7
glimmer=3.02
glimmerhmm=3.0.4
hmmer=2.3.2
hmmer=3.1b2
fasttree=2.1.8
muscle=3.8.31
prodigal=2.6.3
blast=2.2.31
diamond=0.7.12
cd-hit=4.6.6
xz=5.2.4
libxml2=2.9.8
```

4. Install the dependencies.

```
pip install -r requirements.txt
```

5. Ensure that the required Python dependencies are present in the Conda environment.

```
argparse
straight.plugin==1.4.0-post-1
cssselect
pyquery==1.2.9
numpy
biopython>=1.65
helperlibs
pysvg
pyExcelerator
backports.lzma
networkx==1.11
python-louvain==0.5
bcbio-gff>=0.6.2
```

6. Retrieve essential database files, such as protein sequence libraries and pHMMs.

```
python download_databases.py
```

7. Use the main command to predict gene clusters with default settings. Please allocate more memory for big genome size (*see* **Note 6**). Parameters such as search expansion for distant biosynthetic gene loci can be adjusted in plantiSMASH (*see* **Note 7**).

```
python2 run_antismash.py --taxon plants --gff3 species.gff --
clusterblast --knownclusterblast --full-hmmer --outputfolder
~/species/plantismash_output/
```

8. View the prediction results in an HTML web browser for interactive visualization (Fig. 2) and check manually the functional annotations of genes in predicted clusters (*see* **Note 8**).

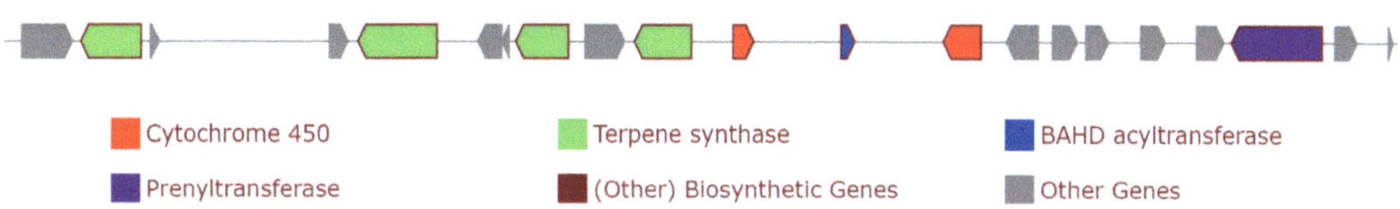

Fig. 2 An example of a biosynthetic gene cluster predicted by plantiSMASH in *H. perforatum* genome

3.2 Syntenic Analysis of Orthologous Gene Clusters Among Phylogenetically Related Species

Similar or identical natural products can be restricted to certain plant lineages, suggesting that these metabolites may be encoded by lineage-specific biosynthetic genes and biosynthetic gene clusters (BGCs). With the growing availability of plant genomes, it is now feasible to investigate the relationship between specific BGCs and their corresponding metabolites. If a BGC is conserved across related species, it strengthens the evidence that this cluster is responsible for the biosynthesis of a particular compound. For example, BGCs encoding polyketide formation for hyperforin in *St. John's wort* or those involved in diterpenoid biosynthesis in Lamiaceae species. Here, we describe a method to explore the conservation and variation of BGCs among phylogenetically related species.

1. Create a new Conda environment for syntenic analysis.

```
conda create -n synteny
conda activate synteny
```

2. Download the JCVI package.

```
pip install jcvi
```

3. Add the LAST [16] dependency to enable the detection and alignment of homologous sequence regions.

```
conda install -c bioconda last
```

4. Generate .bed and .cds genomic files (*see* **Note 9**) from the genome sequence (.fasta) and gene structural annotation files (.gff or .gtf) for each genome (*see* **Note 10**).

```
gffread species1.gff -T -o species1.gtf
gffread species1.gtf -g species1.fasta -x species1.cds
python -m jcvi.formats.gff bed --type=mRNA --key=Name spe-
cies1.gff -o species1.bed
gffread species2.gff -T -o species2.gtf
gffread species2.gtf -g species2.fasta -x species2.cds
python -m jcvi.formats.gff bed --type=mRNA --key=Name spe-
cies2.gff -o species2.bed
```

5. Run the main command to perform syntenic analysis between genomes.

```
python -m jcvi.compara.catalog ortholog species1 species2 --
no_strip_names
```

6. For detailed microsynteny analysis between genomes, the number of aligned regions can be specified using the -iter parameter (*see* **Note 11**).

```
python -m jcvi.compara.synteny mcscan species1.bed species1.
species2.lifted.anchors --iter=1 -o species1.splecies2.i1.
blocks
```

7. Extract BGC-related lines from species1.species2.i1.blocks to create the BGC.blocks file (*see* **Note 12**) and generate a blocks. layout file (*see* **Note 13**) for visualization.

```
# x, y, rotation, ha, va, color, ratio, label
0.5, 0.6, 0, left, center, m, 1, Species1 Chr01
0.5, 0.4, 0, left, center, m, 1, Species2 Sca02
# edges
e, 0, 1
```

8. Merge the two .bed files to create a unified file.

```
cat species1.bed species2.bed > species1_species2.bed
```

9. Visualize the synteny of biosynthetic gene cluster regions between two genomes (Fig. 3).

```
python -m jcvi.graphics.synteny BGC.blocks species1_species2.
bed blocks.layout --format=pdf --shadestyle=curve
```

10. Construct a multi-column BGC.blocks file and visualize the synteny of biosynthetic gene cluster regions among multiple genomes (Fig. 4) and select a proper species as a reference point (*see* **Note 14**).

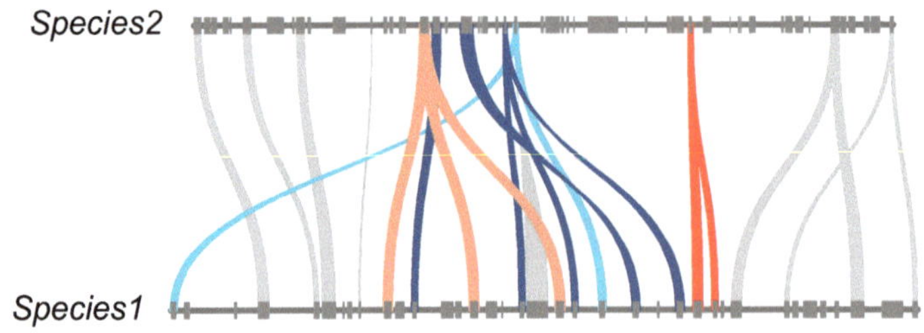

Fig. 3 An example of syntenic analysis between two genomes in the region harboring a biosynthetic gene cluster. Biosynthetic gene clusters in the *H. perforatum* (species 1) and *Mesua ferrea* (species 2) that encode the steps forming polyketide scaffold. Each dark gray box represents a gene locus, and the syntenic BCKDHA, CCL, PKS, and methyltransferase (MT) metabolic genes are highlighted with carnation, light blue, dark blue, and red ribbons, respectively

11. To trace the origin of BGCs within species, the synonymous substitution (Ks) rates between syntenic gene pairs were calculated and converted to million years ago (MYA) using the formula $T = \text{Ks}/2r$ (*see* **Note 15**) to estimate divergence times, with calculations performed by KaKs_Calculator.

```
## download and unpack the package (github.com/lizzhao/
Kaks_Calculator)
tar -zxvf KaKs_Calculator.tar.gz && cd KaKs_Calculator/src &&
make
## align the sequences of syntenic gene pairs
muscle -in seq1_seq2.fa -out seq1_seq2.aln
## calculate the Ks values
KaKs_Calculator -i seq1_seq2.aln -o seq1_seq2.kaks -m YN
```

4 Notes

1. Since plantiSMASH is an extension of the antiSMASH tool, which is based on Python 2 (no longer supported), ensure that the Python version in the Conda environment is not Python 3. For proper compatibility, create a Conda environment with Python 2.7 to avoid potential issues during installation and execution of plantiSMASH.
2. As package versions are frequently updated, the recommended versions for installing dependencies may differ from those listed on the official website. Specifically, plantiSMASH requires both HMMER2 and HMMER3, but installing hmmer=2.3.2 can cause conflicts with HMMER3. Instead, use conda install hmmer2. Additionally, use the following specific versions to avoid compatibility issues:

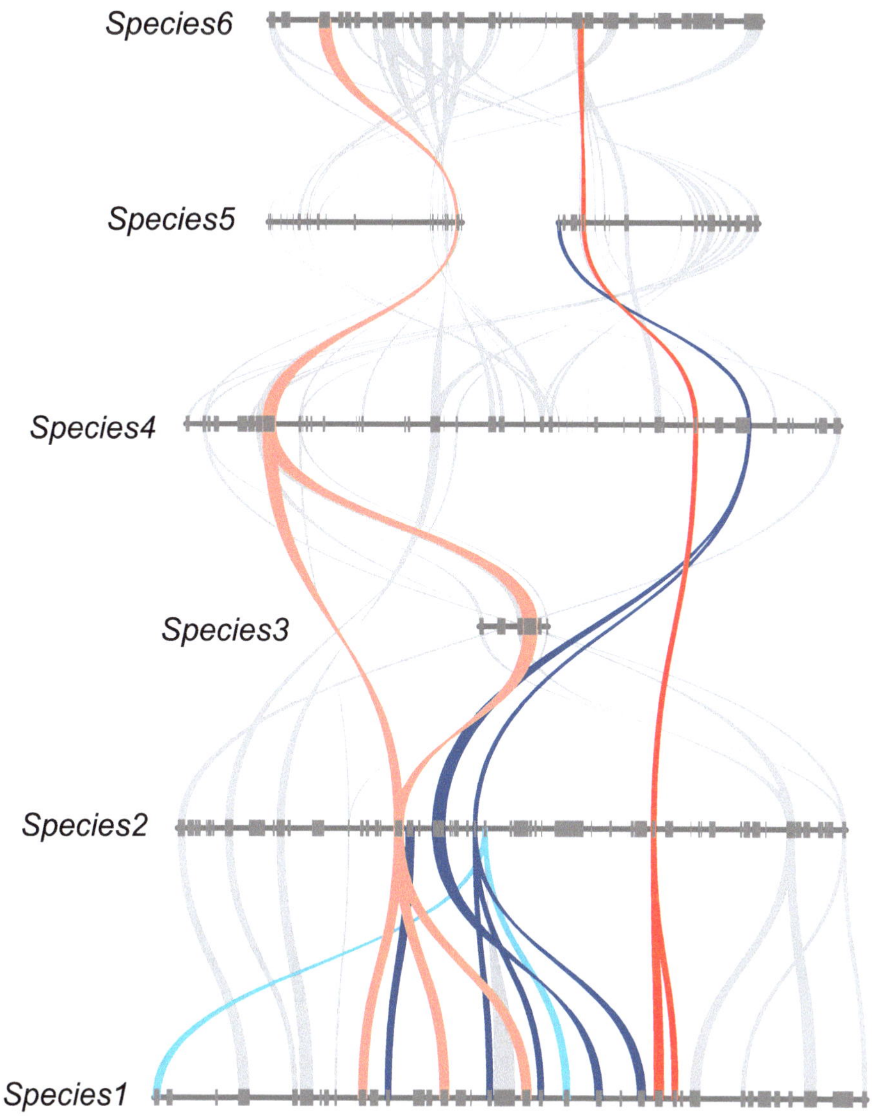

Fig. 4 An example of syntenic analysis across multiple genomes to explore the origin and dynamic evolution of a biosynthetic gene cluster. Synteny analysis of selected plant species within the order Malpighiales reveals conserved genomic regions corresponding to *H. perforatum* BGC1 (species 1). The multi-species comparison of BGC1 highlights two distinct genomic organizations. In *H. perforatum* and *M. ferrea*, the clusters include four gene types—BCKDHA, CCL, PKS, and MT—likely amplified through successive tandem duplications. Conversely, in the rest of the species lacking the ability to produce hyperforin analogs, no BGC is observed that integrates both CCL and PKS. The syntenic metabolic genes BCKDHA, CCL, PKS, and methyltransferase (MT) are represented with ribbons in distinct colors: carnation for BCKDHA, light blue for CCL, dark blue for PKS, and red for MT

- FastTree: 2.1.8 (not 2.1.7)
- Prodigal: 2.6.3 (not 2.6.1)
- DIAMOND: 0.7.12 (not 0.7.10)
- xz: 5.2.4 (not 5.1.1)
- libxml2: 2.9.8 (not 2.9.1).

3. If you encounter this error log, ensure that both MUSCLE and the Java environment are installed on your system. Installing these components should resolve the issue and allow the program to run correctly.

```
ERROR: Failed to locate executable for 'muscle'
ERROR: Failed to locate executable for 'java'
ERROR: Not all prerequisites met.
```

4. If you encounter this error log, ensure that your annotation file (species.gff) adheres to the standard GFF format specifications. Correct formatting is crucial for proper processing and analysis.

```
ERROR: Parsing 'species.gff' failed: ['description', 'RecName:
Full', 'Luminal-binding protein 8']
```

5. If you encounter an error log indicating issues with gene start and end annotations, it means that some entries do not conform to the expected format. To bypass this error, you can temporarily comment out line 750 in SeqFeature.py. This workaround allows the analysis to proceed but should be used cautiously as it might affect data integrity.

```
Traceback (most recent call last)
File "run_antismash.py", line 1829, in <module> main()
File "run_antismash.py", line 631, in main seq_records =
parse_input_sequences(options)
File "run_antismash.py", line 1550, in parse_input_sequences
  det_gene.location = FeatureLocation((gene_start - s_pos),
(e_pos - gene_end))
File "/home/ubuntu/miniconda2/envs/plantismash/lib/python2.7/
site-packages/Bio/SeqFeature.py", line 750, in __init__self.
start))
ValueError: End location (34418) must be greater than or equal
to start location (37084)
```

6. If you encounter a memory error, try allocating more memory to the analysis process.

```
OSError: [Errno 12] Cannot allocate memory
```

7. The --cutoff-multiplier parameter can be adjusted to optimize results, as gene spacing in plant genomes may vary significantly. Experimenting with this parameter allows you to fine-tune the analysis, ensuring more accurate detection of biosynthetic gene clusters under different genomic contexts.
8. Please note that the definition of biosynthetic genes is based on downloaded databases, and in some cases, primary or neo-functionalized metabolic genes might not be detected by plantiSMASH. In such cases, it is recommended to visualize the region using IGV or JBrowse and manually check the annotations to avoid missing any important information.
9. The protein sequence file can be extracted using the -y parameter. However, due to potential mis-annotations in the GFF file—particularly errors in defining gene boundaries—there is a risk of frameshifts in the translation, which can complicate syntenic gene comparisons. To avoid such issues, it is recommended to use CDS (coding sequences) instead, as they are less prone to annotation errors and provide a more accurate representation of gene structures.
10. Transforming a GFF file into a BED file depends on the specific naming conventions used in the GFF file. It is important to set the --type and --key parameters according to the particular features of the GFF file. This ensures that the correct gene or feature types are selected and converted accurately during the transformation process.
11. Typically, for two diploid genomes with closely related phylogeny, most syntenic regions exhibit a 1:1 relationship. However, plant genomes often undergo dynamic changes such as whole genome duplications or large-segmental duplications, resulting in 1:2 or even more complex syntenic relationships. When comparing diploid and polyploid genomes, it is advisable to set the iteration parameter (--iter) to 2 or higher to capture as many syntenic regions as possible.
12. Use plantiSMASH to identify the target gene cluster and note down the corresponding gene IDs. Search these IDs in the syntenic gene pairs file to define the biosynthetic gene cluster region. To ensure comprehensive coverage, extend the selection by 10–15 genes beyond the defined boundaries. Highlight the selected syntenic gene connections in the visualization by

marking lines with "colorcode*" such as "g*". This will display a green ribbon, effectively showing connections and potential genomic rearrangements or translocations that could shift the locus position.

13. In the layout file, you can customize various settings for the final plots, including label colors and the configuration of genomic region positions. This flexibility is particularly useful when working with multiple genomes, as it allows you to organize and visualize the data according to the phylogenetic structure, providing a clearer representation of syntenic relationships.
14. In multi-species syntenic analysis, the order of genome comparison (species1 vs. species2) is crucial. When investigating the origin and evolution of biosynthetic gene clusters, choosing species that are too distantly related can disrupt synteny due to significant differences in genomic architecture. It is beneficial to include an intermediary species that maintains synteny with both distant and closely related species. For example, *Andrographis paniculata* within Lamiales shows good synteny with both tomato and other Lamiaceae species, making it an ideal reference point for comparative analysis.
15. The evolutionary rate of synonymous substitutions per site per year (r) can vary between different species or genomes. For angiosperms, the average synonymous substitution rate is typically 6.5×10^{-9} substitutions per site per year, meaning that approximately 6.5×10^{-9} synonymous substitutions occur at each nucleotide site per year.

Acknowledgments

This work was supported by the Strategic Priority Research Program of the Chinese Academy of Sciences (Grant XDB2702020).

References

1. Nützmann HW, Huang A, Osbourn A (2016) Plant metabolic clusters – from genetics to genomics. New Phytol 211:771–789. https://doi.org/10.1111/nph.13981
2. Polturak G, Liu Z, Osbourn A (2022) New and emerging concepts in the evolution and function of plant biosynthetic gene clusters. Curr Opin Green Sustain Chem 33:100568. https://doi.org/10.1016/j.cogsc.2021.100568
3. Medema MH, Osbourn A (2016) Computational genomic identification and functional reconstitution of plant natural product biosynthetic pathways. Nat Prod Rep 33:951–962. https://doi.org/10.1039/c6np00035e
4. Méteignier LV, Nützmann HW, Papon N et al (2023) Emerging mechanistic insights into the regulation of specialized metabolism in plants. Nat Plants 9:22–30. https://doi.org/10.1038/s41477-022-01288-7
5. Li H, Wu S, Lin R et al (2023) The genomes of medicinal skullcaps reveal the polyphyletic origins of clerodane diterpene biosynthesis in the family Lamiaceae. Mol Plant 16:549–570. https://doi.org/10.1016/j.molp.2023.01.006

6. Wu S, Malaco Morotti AL, Wang S et al (2022) Convergent gene clusters underpin hyperforin biosynthesis in St John's wort. New Phytol 235:646–661. https://doi.org/10.1111/nph.18138
7. Wu S, Morotti ALM, Yang J et al (2024) Single-cell RNA sequencing facilitates the elucidation of the complete biosynthesis of the antidepressant hyperforin in St. John's wort. Mol Plant 17:1439–1457. https://doi.org/10.1016/J.MOLP.2024.08.003
8. Zhao Q, Yang J, Cui MY et al (2019) The reference genome sequence of Scutellaria baicalensis provides insights into the evolution of wogonin biosynthesis. Mol Plant 12:935–950. https://doi.org/10.1016/j.molp.2019.04.002
9. Polturak G, Dippe M, Stephenson MJ et al (2022) Pathogen-induced biosynthetic pathways encode defense-related molecules in bread wheat. Proc Natl Acad Sci USA 119. https://doi.org/10.1073/PNAS.2123299119/-/DCSUPPLEMENTAL
10. Martin LBB, Kikuchi S, Rejzek M et al (2024) Complete biosynthesis of the potent vaccine adjuvant QS-21. Nat Chem Biol. https://doi.org/10.1038/S41589-023-01538-5
11. Kautsar SA, Suarez Duran HG, Blin K et al (2017) plantiSMASH: automated identification, annotation and expression analysis of plant biosynthetic gene clusters. Nucleic Acids Res 45:W55–w63. https://doi.org/10.1093/nar/gkx305
12. Kautsar SA, Suarez Duran HG, Medema MH (2018) Genomic identification and analysis of specialized metabolite biosynthetic gene clusters in plants using PlantiSMASH. Methods Mol Biol 1795:173–188. https://doi.org/10.1007/978-1-4939-7874-8_15
13. Tang H, Bowers JE, Wang X et al (2008) Synteny and collinearity in plant genomes. Science. (1979 320:486–488
14. Tang H, Krishnakumar V, Zeng X et al (2024) JCVI: a versatile toolkit for comparative genomics analysis. iMeta 3:e211. https://doi.org/10.1002/IMT2.211
15. Wang D, Zhang Y, Zhang Z et al (2010) KaKs_Calculator 2.0: a toolkit incorporating gamma-series methods and sliding window strategies. Genomics Proteomics Bioinformatics 8:77–80. https://doi.org/10.1016/S1672-0229(10)60008-3
16. Kiełbasa SM, Wan R, Sato K et al (2011) Adaptive seeds tame genomic sequence comparison. Genome Res 21:487–493. https://doi.org/10.1101/GR.113985.110

Chapter 3

Mining Metabolic Diversities Across Genomic Neighborhoods in Plants

Xuan Zhou and Zhenhua Liu

Abstract

There is increasing evidence that genes in plant genomes are not randomly distributed. Notable examples are the recently identified biosynthetic gene clusters (BGCs), which encode various specialized metabolic pathways. The evolution of BGCs in plants is highly debated, mainly due to the complex interplay of gene duplication, cryptic genomic translocation, and selective pressures that drive their formation and diversification. We previously adopted genomic neighborhood (GN) analysis to identify and interpret evolutionary events and trajectories for BGCs specifically evolved in Brassicaceae. Here, we present step-by-step methods for conducting GN association analysis to identify significantly clustered genes, utilizing high-quality genomes from grass species as working materials. This pipeline provides a framework for understanding the organization (assessed by content similarity) and evolution (evaluated by sequence similarity) of plant BGCs, offering insights into the mechanisms underlying the distribution of genes involved in specialized metabolism.

Key words Genomic neighborhood association analysis, Biosynthetic gene clusters, Metabolic diversity

1 Introduction

Similar to bacterial operons, certain metabolic pathway genes in plant genomes are tightly clustered, forming biosynthetic gene clusters (BGCs) [1, 2]. However, plant BGCs differ from operons in several ways. Each BGC gene has its own promoter, yet often maintains co-expression with neighboring genes. Plant BGCs likely evolved through gene duplication, followed by genomic translocation and neo-functionalization, whereas operon diversification mainly relies on horizontal gene transfer [3–5]. Since plant BGC genes are not co-transcribed under a single promoter like operons, identifying them can be challenging due to the lack of clear boundaries within genomic neighborhoods containing metabolic genes. Nevertheless, bioinformatic tools have been developed to mine plant BGCs [6, 7], drawing inspiration from microbial BGC

Hans-Wilhelm Nützmann and Ramesha Thimmappa (eds.), *Metabolic Gene Clusters: Methods and Protocols*,
Methods in Molecular Biology, vol. 3054, https://doi.org/10.1007/978-1-0716-5384-5_3,

studies. These tools feature known scaffolding genes, which likely encode committed steps or branch points in metabolic pathways [8], and use them as anchor genes to search for nearby flanking genes within a defined genomic neighborhood (GN). When multiple (usually more than two) decorating genes (encoding enzymes that are able to modify the scaffold) are found in the GNs, candidate BGCs are identified. Although these methods have successfully identified, on average, hundreds of BGCs in various plant genomes—including functionally characterized ones—many of the identified genes either do not exhibit co-expression patterns or are not functionally related. Additionally, they do not provide evolutionary information, even for characterized BGCs. So far, most of the reported BGCs are narrowly distributed in specific plant families or lineages, and little is known about their evolutionary history (birth, diversification, and death), often due to the lack of comparable homologous gene clusters [9] (Fig. 1).

To facilitate accurate identification of plant BGCs and uncover distinct evolutionary trajectories, Liu et al. adopted an all-vs-all comparison to explore the diversity of triterpene BGCs across 13 high-quality Brassicaceae genomes [10]. Plant triterpenes are derived from the ubiquitous precursor, 2,3-oxidosqualene, which is folded into more than 100 distinct backbones by scaffolding enzymes known as oxidosqualene cyclases (OSCs) [11]. Liu et al. identified 163 OSC genes from Brassicaceae genomes and categorized them into three distinct phylogenetic clades: clade I, clade II, and the basal clade. They then defined OSC-centric GNs by extending either five or ten genes next to the OSC gene for each identified clade. To determine if the flanking genes are significantly enriched in the OSC GNs, they applied a hypergeometric test, comparing the occurrence of specific genes in the GN with their probability in the entire genome. This analysis identified candidate genes that may be functionally associated with OSCs, including those encoding non-enzymatic proteins. To interpret the evolutionary relationships between these OSC GNs, the authors conducted an all-vs-all comparison of both GN content and gene sequence similarities, using the tools developed in BIG-SCAPE [12]. These comprehensive analyses produced a heatmap illustrating the conservation and diversification of triterpene BGCs across a wide range of Brassicaceae species.

Here, we adopt and adapt the GN analysis, previously applied to OSC GNs in the Brassicaceae, to investigate terpene synthase (TPS)-involved GNs in the grass family (Poaceae). Terpene synthases are key enzymes responsible for the biosynthesis of a wide variety of terpenes, including monoterpenes (C10), sesquiterpenes (C15), and diterpenes (C20) [13]. We present a detailed methodology for identifying TPS-centric GNs across representative grass species with high-quality genomes. Our approach combines

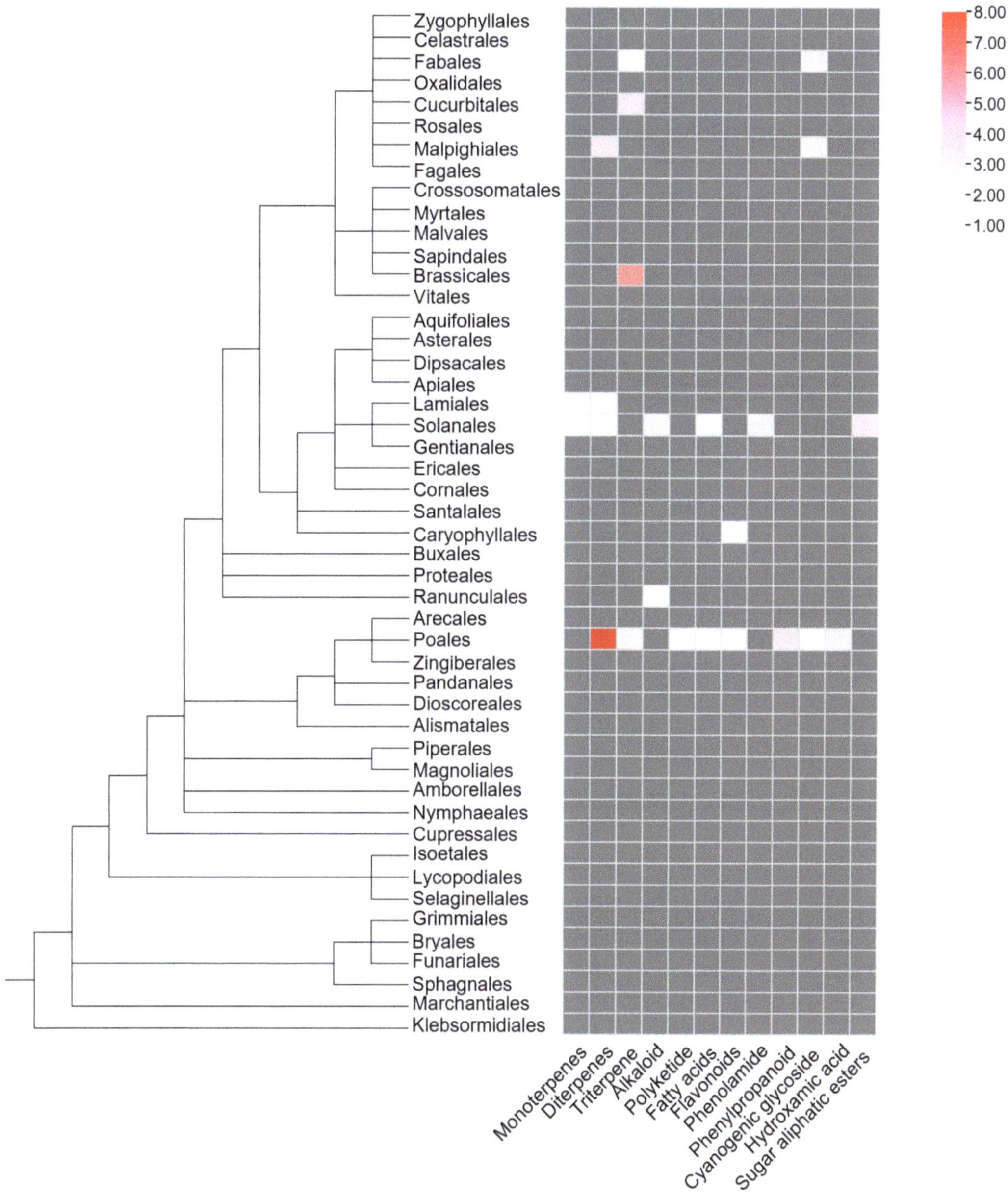

Fig. 1 The distribution of the reported BGCs in plant orders. To date, BGCs have been identified in multiple plant orders, with notable occurrences in Poales and Solanales. Specifically, diterpene BGCs are prevalent in Poales, while triterpene BGCs are common in Brassicales. The color reflects the number of reported BGCs in the plant families/orders. This summary includes data up to January 2025

GN association analysis with comprehensive all-vs-all comparisons, demonstrating the utility of these tools in elucidating the conservation and diversification of TPS biosynthetic gene clusters (BGCs) within the grass family (Fig. 2).

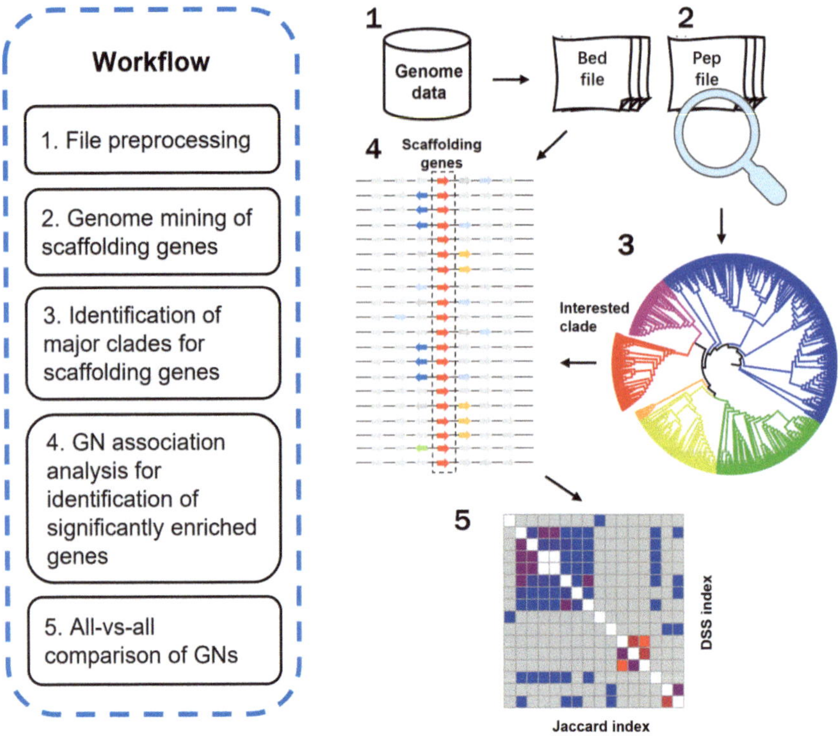

Fig. 2 The workflow of the GN analysis

2 Materials

The grass family includes a large number of important crops (e.g. rice, maize, and wheat) that possess various types of reported BGCs (Fig. 1). Representative high-quality genomes (chromosome-assembly), along with their corresponding annotation files (i.e., gff3 file), for plants in the Poaceae family were downloaded from the public database (Table 1).

3 Methods

3.1 Dependencies and Pipeline Input Files

1. The described pipeline is recommended to be executed on a Linux system for optimal performance. Data processing is performed using Python (version 3.10), and visualization is achieved with R (version 4.2). All software used in this study are listed in Table 2.
2. Two input files, including representative protein sequences (pep files) and chromosomal position of each gene (bed files), are prepared by retrieving information from the annotation file. Notably, the gene names in both files must be consistent to ensure accurate analysis.

Table 1
Summary of examined grass plants in the study

Organism name	Abbreviation	Subfamily	Database
Sorghum bicolor	Sbi	Andropogoneae	NCBI
Zea mays	Zma	Andropogoneae	Phytozome
Brachypodium distachyon	Bdi	Brachypoideae	Phytozome
Oropetium thomaeum	Oth	Cynodonteae	Phytozome
Eragrostis curvula	Ecu	Eragrostideae	NCBI
Oryza sativa Japonica Group	Osa	Oryzeae	Phytozome
Setaria italica	Sit	Paniceae	NCBI
Setaria viridis	Svi	Paniceae	NCBI
Paspalum vaginatum	Pva	Paspaleae	NCBI
Pharus latifolius	Pla	Pharoideae	Phytozome
Lolium perenne	Lpe	Poeae	NCBI
Avena strigosa	Oat	Poeae	http://db.ncgr.ac.cn/oat/index.php
Triticum aestivum	Tae	Triticeae	NCBI
Hordeum vulgare	Hvu	Triticeae	Phytozome
Lolium rigidum	Lri	Poeae	NCBI
Triticum urartu	Tur	Triticeae	NCBI
Aegilops tauschii subsp. strangulata	Ata	Triticeae	NCBI
Triticum dicoccoides	Tdi	Triticeae	NCBI
Triticum turgidum subsp. durum	Ttu	Triticeae	NCBI

3.2 Identification of the GN Scaffolding Genes

Here, we define genomic neighborhoods (GNs) as genomic blocks centered around genes encoding scaffolding enzymes for specialized metabolic pathways [9, 10]. We demonstrate the pipelines by working on the terpene synthases (TPS) GNs in the grass family.

1. To identify these TPS GNs, systematic identification of TPS anchor genes is essential. HMMER3 [14] was used to find TPS homologs.

```
hmmsearch --domtblout PF01397.out --cut_tc PF01397.hmm protein.fasta
```

Table 2
The required software for the pipeline

Tools/database	Purpose
Python v3.10 pandas-2.2.3 scipy-1.10.0 numpy-1.23.5 statsmodels-0.14.4	Python scripts
R version 4.2.3 pheatmap-1.0.12 ggplot2-3.3.6 gggenes-0.5.1 dplyr-1.0.9	Scripts visualization
HMMER-3.4	Annotating protein sequence
Pfam-A database	Annotating protein sequence
MUSCLE-5.1	Multiple sequences alignment
FastTree-2.1.11	Phylogenetic tree construction

`hmmsearch --domtblout PF03936.out --cut_tc PF03936.hmm protein.fasta` # TPS homologs were searched using two PFAM (https://pfam.xfam.org) profiles, PF01397 and PF03936, which target the N-terminal and C-terminal of TPS, respectively.

2. All TPS proteins identified by HMMER3 were aligned with functionally characterized marker genes using MUSCLE [15] to generate a multiple sequence alignment (MSA).

`cat PF01397.out PF03936.out | grep ^[^#] | awk '{print $1,$3}'> TPS.list` # Retrieve the name and length of every candidate TPS protein.

`awk '$2>500' TPS.list | awk '$2<900' > TPS-500-900.list` # Sequences that are too long (> 900aa) or too short (< 500aa) for a typical TPS were removed.

`python get_fa_from_id.py protein.fasta TPS.fasta TPS-500-900.list` # Extract sequences according to the sequence_id. Up to this point, 842 TPS proteins were identified from the 19 selected species.

`cat TPS.fasta tps_ref.fa > all.fa` # Add marker proteins that are functionally characterized to the fasta file, so as to have support for the identification of clades that are likely to have evolved from common ancestors in the phylogenetic tree.

`muscle -in all.fa -out all.msa` # MSA output.

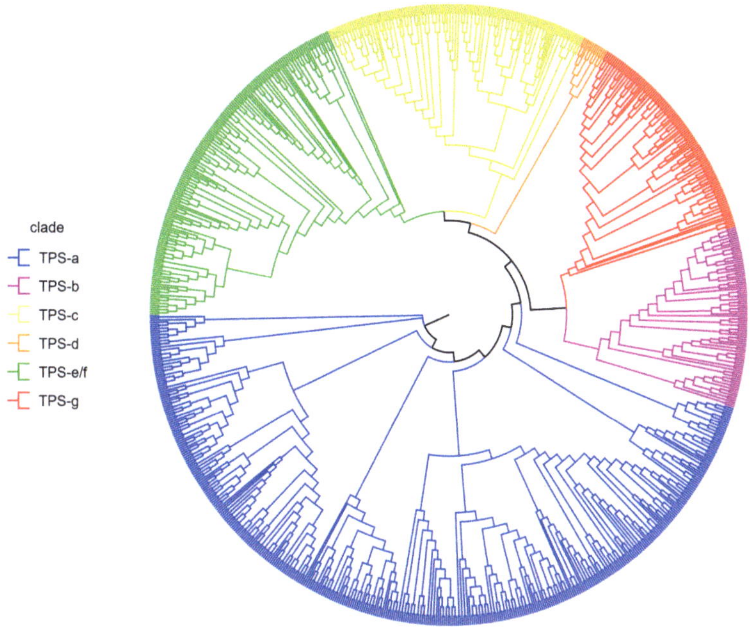

Fig. 3 The phylogenetic tree of identified TPS genes from the Poaceae family. Different colors represent different TPS subfamilies

3. The MSA was used to build a phylogenetic tree using FastTree 2.1 [16] with standard parameters.

`FastTree all.msa > all_tree #` A maximum-likelihood phylogenetic tree was reconstructed.

4. One of the main reasons for classifying GNs using scaffolding genes is that these genes are likely to have evolved early in evolutionary history. Consequently, these foundational genes are typically well-conserved and less impacted by dynamic sequence evolution. Therefore, phylogenetic analysis of scaffolding genes can provide a more resolved tree topology compared to decorating genes, which are more diverse and often more recently evolved. For the generated TPS protein tree, six major clades were resolved, consistent with previous studies that classify TPSs into six major subfamilies (Fig. 3) [13].

3.3 Genomic Neighborhoods Association Analysis

1. A simple script was developed to isolate genomic neighborhoods (GNs) based on a user-specified number of flanking genes. For example, considering GNs with X genes flanking the anchor genes (marker loci) results in GNs consisting of 2X + 1 genes (*see* **Note 1**). Two exceptions may occur: (1) if marker loci are located in a scaffold with fewer than 2X + 1 genes, smaller GNs are generated, and (2) if tandemly

duplicated *TPSs* are present, overlapping GNs are merged into a larger GN containing more than 2X + 1 genes. An effective GN includes about 5–10 genes flanking the scaffolding gene, based on the fact that most studied specialized metabolic pathways fall within this gene size range [17, 18].

`cat *.bed > all_bed.txt` # Merge the gene chromosomal position information from all genomes into a single file.

`sh gene_neighbor.sh all_bed.txt TPS-a.txt 5 > TPS_a.log` # This shell script extracts anchor genes within genomic neighborhoods (GNs). Here, each GN spans the anchor locus plus five upstream and five downstream genes. The TPS_a.log file flags any GN whose gene count deviates from the expected 2X + 1 size.

`sh neighbor.sh` # Execute this step in the same directory as the previous one.

2. Next, we assessed the enrichment of genes (represented by PFAM domains) in the TPS GNs for each TPS clade, which are likely to share the most recent common ancestors. This was done by comparing them to the rest of the genomes using a hypergeometric test (one-tailed exact Fisher's test) with the Bonferroni correction for multiple comparisons (*see* **Note 2**).

`hmmscan --domtblout all.hmmout --cut_tc --cpu 5 Pfam-A.hmm all.pep` # After retrieving all genes within TPS genomic neighborhoods across the selected genomes, their Pfam domain architectures were predicted with hmmscan using the --cut_tc threshold.

`python hypergeom.py all_bed.txt all.bed all.hmmout TPS_a.result` # The all.bed file is produced by neighbor.sh. Our custom Python script performs hypergeometric tests via fisher_exact and corrects for multiple comparisons with multipletests.

$$P(X = k) = f(k; N, K, n) = \frac{\binom{K}{k}\binom{N-K}{n-k}}{\binom{N}{n}}$$

k: this_pfam_clustered *K*: all_clustered_pfams

n: this_pfam_total *N*: all_total_pfams

For instance, in the TPS-b genomic neighborhoods (GNs), *K* and *N* are 428 and 598,269, respectively. The *k* and *n* values for each targeted domain are listed in Table 3. The hypergeometric test identifies PFAM domains that are significantly enriched in the GNs, with *p*-values provided in Table 3. In Fig. 4, the significantly enriched domains are labeled within the GNs, organized according to the phylogeny of TPS-b.

Table 3
Domains significantly associated with the Poaceae terpene synthase (TPS) from the TPS-b clade

Pfam domain	*P* < 0.01 (Fisher, bfr)	*k*	*n*
p450	7.0075E-51	71	8917
CDT1_C	1.2914E-21	10	33
ClpS	5.1095E-19	9	32
Lipase_GDSL	3.9313E-16	25	3478
H2TH	4.0146E-10	5	16
CDT1	1.7618E-06	4	29
SPARK	3.9925E-06	5	111
Ribosomal_S25	1.6882E-05	4	68
zf-C3HC4_2	2.0155E-05	9	1101
Amidase	2.2904E-05	6	336
Fapy_DNA_glyco	3.5462E-05	3	13
THH1_TOM1-3_dom	2.3031E-04	4	125
TTI1	2.6120E-04	3	24
Nucleotid_trans	5.8766E-03	5	553
Fpg-like_C	7.6381E-03	2	16

3.4 All-vs-All Comparison of GNs

1. To examine the gene content and sequence similarity of the identified TPS GNs, we employed BiG-SCAPE to evaluate two criteria: domain composition similarity, measured by the Jaccard index of identified protein domains, and domain sequence similarity (DSS), assessed by averaging the sequence similarity of shared domains, with a penalty applied when a GN contains additional copies of a protein domain not present in the other [12].

`python GN_comp.py TPS-b.txt bed_dir all.pep all.hmmout Pfam-A.hmm` # GN similarities were computed with the GN_comp.py script.

`Rscript heatmap.r` # The heatmap.r script employs the pheatmap package to visualize both the Jaccard index and domain-sequence similarity.

The heatmap, organized according to the TPS-b phylogeny, illustrates the conservation and diversification of TPS-b GNs within the grass family (Fig. 5). By examining the scaffolding TPS genes, one can assess the content and sequence similarity of a specific GN

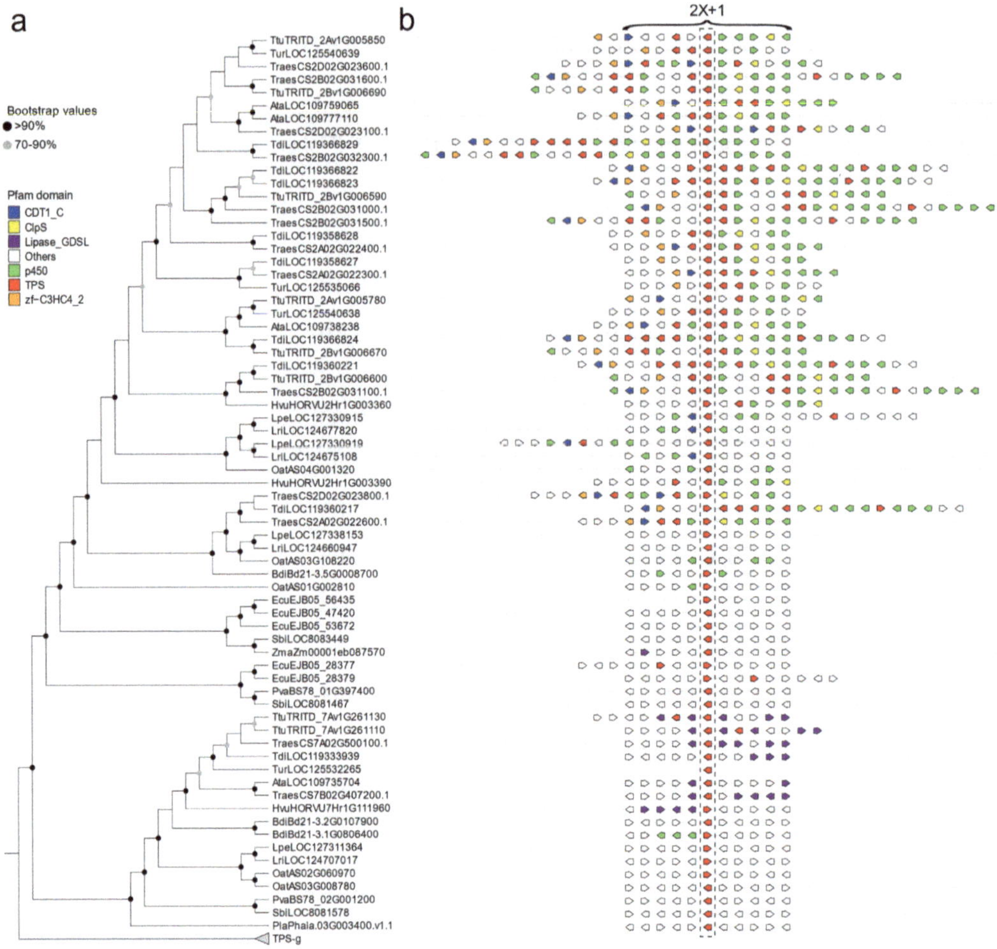

Fig. 4 Domains that are significantly enriched in TPS-b clade GNs. (**a**) Maximum-likelihood tree of TPS-b protein sequences. (**b**) TPS-b GNs. The genes encoding the TPS in each tree leaf are positioned in the middle of GN. Arrows denote the strand directionality of genes. Domains that are significantly enriched and appeared more than 9 times in the TPS-b GNs (Table 3) are colored

relative to other TPS-b GNs. For example, the region labeled "hot region 1" in Fig. 5 encompasses TPS-b GNs confined to the Poeae lineage, indicating their recent evolutionary origin. Additionally, the comparable Jaccard and Domain Sequence Similarity (DSS) indices in this region suggest that these GNs maintain conserved biochemical functions. Conversely, "hot region 2" exhibits a higher Jaccard index (indicating similar GN content) compared to the DSS index (indicating sequence diversification), implying rapid evolution among these similar GNs.

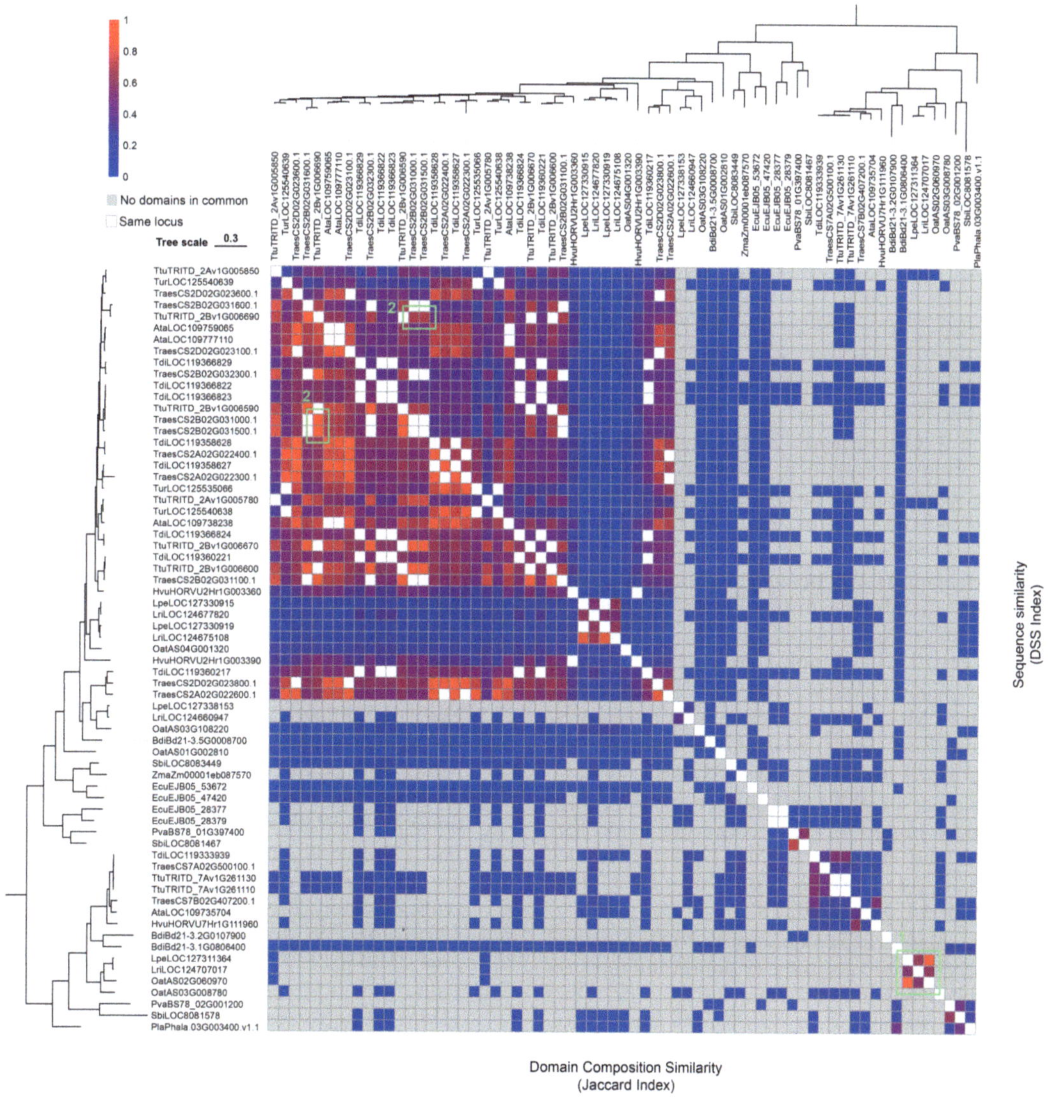

Fig. 5 Similarity comparison of TPS-b GNs in the grass family. The heatmap is ordered based on the phylogenetic relationship of TPS-b genes. The white diagonal line separates the two indices for the measurement of domain composition and sequence similarity, respectively. Gray indicates that there are no shared domains between the compared GN pairs. Red and blue indicate high and low similarity between GNs, respectively

4 Notes

1. We have provided a script (see GitHub) to adjust the GN size.
2. The GN association analysis can also be applied to a single genome. Type I error may increase for analysis in multiple genomes, due to the complex evolution of plant genomes.

Code Availability

Commands and scripts used in this study are available at https://github.com/XuanZhou001/GN_association.

Acknowledgment

This work was supported by the National Key R&D Program of China (2022YFC2602000).

References

1. Field B, Osbourn AE (2008) Metabolic diversification—independent assembly of operon-like gene clusters in different plants. Science 320:543–547. https://doi.org/10.1126/science.1154990
2. Nützmann H-W, Huang A, Osbourn A (2016) Plant metabolic clusters—from genetics to genomics. New Phytol 211:771–789. https://doi.org/10.1111/nph.13981
3. Nützmann H-W, Scazzocchio C, Osbourn A (2018) Metabolic gene clusters in eukaryotes. Annu Rev Genet 52:159–183. https://doi.org/10.1146/annurev-genet-120417-031237
4. Lawrence J (1999) Selfish operons: the evolutionary impact of gene clustering in prokaryotes and eukaryotes. Curr Opin Genet Dev 9:642–648. https://doi.org/10.1016/S0959-437X(99)00025-8
5. Polturak G, Liu Z, Osbourn A (2022) New and emerging concepts in the evolution and function of plant biosynthetic gene clusters. Curr Opin Green Sustain Chem 33:100568. https://doi.org/10.1016/j.cogsc.2021.100568
6. Kautsar SA, Suarez Duran HG, Blin K et al (2017) plantiSMASH: automated identification, annotation and expression analysis of plant biosynthetic gene clusters. Nucleic Acids Res 45:W55–W63. https://doi.org/10.1093/nar/gkx305
7. Töpfer N, Fuchs L-M, Aharoni A (2017) The PhytoClust tool for metabolic gene clusters discovery in plant genomes. Nucleic Acids Res 45:7049–7063. https://doi.org/10.1093/nar/gkx404
8. Ji W, Osbourn A, Liu Z (2024) Understanding metabolic diversification in plants : branch-points in the evolution of specialized metabolism. Philos Trans R Soc B Biol Sci 379:1914. https://doi.org/10.1098/rstb.2023.0359
9. Zhou X, Liu Z (2022) Unlocking plant metabolic diversity: a (pan)-genomic view. Plant Commun 3:100300. https://doi.org/10.1016/j.xplc.2022.100300
10. Liu Z, Suarez Duran HG, Harnvanichvech Y et al (2020) Drivers of metabolic diversification: how dynamic genomic neighbourhoods generate new biosynthetic pathways in the Brassicaceae. New Phytol 227:1109–1123. https://doi.org/10.1111/nph.16338
11. Xu R, Fazio GC, Matsuda SPT (2004) On the origins of triterpenoid skeletal diversity. Phytochemistry 65:261–291. https://doi.org/10.1016/j.phytochem.2003.11.014
12. Navarro-Muñoz JC, Sélem-Mojica N, Mullowney MW et al (2020) A computational framework to explore large-scale biosynthetic diversity. Nat Chem Biol 16:60–68. https://doi.org/10.1038/s41589-019-0400-9
13. Jia Q, Brown R, Köllner TG et al (2022) Origin and early evolution of the plant terpene synthase family. Proc Natl Acad Sci USA 119: e2100361119. https://doi.org/10.1073/pnas.2100361119
14. Finn RD, Clements J, Eddy SR (2011) HMMER web server: interactive sequence similarity searching. Nucleic Acids Res 39:W29–W37. https://doi.org/10.1093/nar/gkr367
15. Edgar RC (2004) MUSCLE: multiple sequence alignment with high accuracy and high throughput. Nucleic Acids Res 32:1792–1797. https://doi.org/10.1093/nar/gkh340
16. Price MN, Dehal PS, Arkin AP (2010) FastTree 2–approximately maximum-likelihood trees for large alignments. PLoS One 5:e9490. https://doi.org/10.1371/journal.pone.0009490
17. Hawkins C, Xue B, Yasmin F et al (2025) Plant Metabolic Network 16: expansion of underrepresented plant groups and experimentally supported enzyme data. Nucleic Acids Res 53: D1606–D1613. https://doi.org/10.1093/nar/gkae991
18. Zdouc MM, Blin K, Louwen NLL et al (2025) MIBiG 4.0: advancing biosynthetic gene cluster curation through global collaboration. Nucleic Acids Res 53:D678–D690. https://doi.org/10.1093/nar/gkae1115

Chapter 4

Identification of Stress-Induced Biosynthetic Gene Clusters in Plants

Amir Ronen, Boaz Negin, Michal Rosenthal, and Guy Polturak

Abstract

Phytoalexins are specialized metabolites that play a crucial role in plant defense response. Similarly to other plant specialized metabolites, their biosynthetic pathways are in some cases encoded by co-localized genes that form biosynthetic gene clusters (BGCs). Discovery of BGCs responsible for phytoalexin production can significantly facilitate delineation of their biosynthesis and function. Here we describe an experimental approach for systematic identification of stress-induced BGCs, potentially involved in phytoalexin biosynthesis. Plants are treated with methyl jasmonate (MeJA) to induce stress-responsive metabolic pathways, and RNA sequencing (RNA-seq) of treated plants is applied to identify genes upregulated in response to MeJA treatment. The resulting transcriptomic data is next integrated with BGC prediction tools to identify co-localized genes, together leading to the identification of stress-induced BGC candidates. This approach provides a pipeline that serves to facilitate the discovery of novel phytoalexin biosynthetic pathways, thus providing insight into defense-related plant specialized metabolism and its underlying genetic basis.

Key words Phytoalexins, Biosynthetic gene clusters, Plant defense, Specialized metabolism, Methyl jasmonate

1 Introduction

The occurrence of biosynthetic gene clusters (BGCs) in plant genomes has been pivotal for the elucidation of biosynthetic pathways of various plant specialized metabolites [1]. "Genome mining" computational tools have therefore been developed for the identification of BGC candidates from plant genomic data [2–4]. These tools typically identify dozens of candidate clusters per genome. However, it is not possible, using genomic data alone, to anticipate whether these gene clusters are functional, leading to the synthesis of predicted products. Furthermore, even a functional cluster may be expressed in a tissue- or condition-specific manner. In particular, BGCs encoding the production of phytoalexins (antimicrobial small molecules produced in response to biotic stress) are often markedly induced following pathogen infection or related biotic

Hans-Wilhelm Nützmann and Ramesha Thimmappa (eds.), *Metabolic Gene Clusters: Methods and Protocols*,
Methods in Molecular Biology, vol. 3054, https://doi.org/10.1007/978-1-0716-5384-5_4,

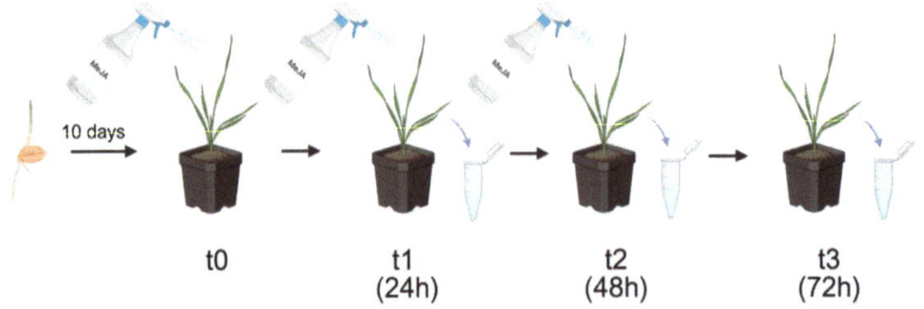

Fig. 1 Treatment and sampling of plants. Wheat plants are sprayed with a methyl jasmonate (MeJA) solution, or a mock solution, 10 days post germination (t0), and repeatedly sprayed 24 and 48 h after the first treatment. Plants are sampled 24, 48, and 72 h after the first treatment for RNA extraction and purification. (Created in BioRender. https:/BioRender.com/syhscwu)

stress cues [5–7]. Complementation of BGC genome mining with transcriptomic data from stress-induced plants is thus highly useful for identifying BGCs that potentially produce phytoalexins, as part of the plant defense response. Here we describe a method for generating transcriptomic data from stress-induced plants, and its integration with gene clustering information, as a means of identifying stress-induced BGC candidates. We outline steps for the treatment of plants with a stress-response elicitor, RNA extraction for RNA-sequencing, and the analysis of the resulting RNA-Seq data.

Generation of cDNA for quantitative real-time PCR (qRT-PCR) analysis is also described, as this is useful for a preliminary assessment of gene induction prior to RNA sequencing (i.e., by analyzing expression of known MeJA-induced genes). In the method described here we apply the chemical elicitor methyl jasmonate (MeJA) on wheat, but a similar approach can be applied to other plant species, and using different elicitors (e.g., $CuCl_2$, salicylic acid, chitin, flagellin). Considering that phytoalexin production is frequently tissue-specific, sampling of more than one tissue type is advised (e.g., shoots and roots). It is also advisable to sample plant material for RNA extraction in 2–3 timepoints post treatment, so that samples exhibiting optimal induction can be determined by qRT-PCR and selected for sequencing and downstream analysis. Here we sample at 24, 48, and 72 h post first treatment (Fig. 1), but other timepoints may be considered (e.g., 6, 12 h post treatment), as response times for gene induction vary considerably. Consulting literature for previously reported experiments with the same elicitor and plant species used in the experiment is therefore recommended.

2 Materials

2.1 Plant Material

1. Seeds of *Triticum aestivum*, cv. Chinese Spring (bread wheat).

2.2 Sterilization Solutions

1. Ethanol 70%: Add 35 mL absolute ethanol to a 50 mL plastic tube and make up to 50 mL with water (*see* **Note 1**). Store at room temperature (RT).
2. 1.5% sodium hypochlorite (bleach): Add 12.5 mL from sodium hypochlorite stock solution (6%) to a 50 mL plastic tube and make up to 50 mL with water (*see* **Note 1**). Store at RT.

2.3 Spraying Solutions

1. 200 μM methyl jasmonate (MeJA) solution: Add 22.9 μL MeJA from a stock solution (1.03 g/mL) into 1 mL absolute ethanol (*see* **Note 2**). Next, slowly add the MeJA/ethanol solution to 500 mL water in a glass or plastic beaker, while stirring. Add 50 μL Tween 20 surfactant (final concentration 0.01%) to the solution while stirring. The solution must be freshly prepared just before use.
2. Mock solution; To 0.5 L water in glass or plastic beaker, add 1 mL absolute ethanol while stirring. Then add 50 μL Tween 20 (final concentration 0.01%) to the solution while stirring. The solution must be prepared just before use.

2.4 Plant Growth and Sampling

1. 50 mL polypropylene tubes.
2. Scalpel.
3. Tweezers.
4. Petri dish (90 mm).
5. Filter paper (e.g., Whatman No.3 90 mm circle), autoclaved.
6. Plastic pots (7L × 7W × 6D cm).
7. Plastic trays (54L × 28W × 6D cm).
8. Clear plastic cover for the plastic tray (54L × 28W × 18D cm).
9. Soil suitable for small pots, containing peat and controlled-release fertilizer.
10. Growth chambers (16 h light, 8 h dark, 20 °C during light and 15 °C during dark).

2.5 RNA Extraction and Purification

1. TRIzol solution.
2. Liquid nitrogen.
3. Mortar and pestle.
4. Chloroform.
5. Isopropanol.
6. DEPC-treated water/RNase-free water.

2.6 cDNA Synthesis and Quantitative Real-Time PCR (qRT-PCR)

1. cDNA reverse transcription kit.
2. DNase I kit.
3. RNA inhibitor.
4. SYBR Green I fluorescent dye.
5. Real-time PCR tubes/plate.
6. Real-Time PCR instrument.

3 Methods

Carry out all procedures at room temperature (RT), unless otherwise specified.

3.1 Seed Sterilization and Germination

1. Place approximately 100 wheat seeds in a 50 mL polypropylene tube.
2. Add to the seeds 20 mL of 70% ethanol and rotate the tube slowly (10 rpm) for 1 min.
3. Discard the ethanol by careful decanting (*see* **Note 3**).
4. Add 20 mL of 1.5% sodium hypochlorite to the tube with the seeds and rotate the tube slowly (10 rpm) for 10 min.
5. Discard the sodium hypochlorite solution by careful decanting (*see* **Note 3**).
6. Wash the seeds with 20 mL water by inverting the tube 4–6 times and discarding the water.
7. Repeat **step 6** two more times.
8. In a biological hood, place autoclaved filter paper in a standard-sized Petri dish cover. Wet the paper with 1–2 mL of autoclave-sterilized water.
9. Spread 20–30 seeds on top of each wet filter paper (*see* **Note 4**).
10. Cover and seal the plates with parafilm. Incubate in the dark at 4 °C for two nights (*see* **Note 5**).
11. For germination, place the plates in a plant growth chamber for 2–5 days, until the root and hypocotyl emerge (*see* **Note 6**).
12. Prepare 72 plastic pots (7L × 7W × 6D cm) with water-saturated soil in plastic trays (54L × 28W × 6D cm). Plants are divided to comprise two treatments (MeJA/mock), which are sampled at three timepoints (24, 48, 72 h post first treatment), in three biological replicates, each containing four pooled individual plants. Plants for MeJA or for mock treatments are placed on separate trays.

13. Transfer one seedling to each pot, placing it in a small cavity in the soil and covering gently, with the root facing down and hypocotyl facing up, towards the light source.
14. Place the trays in the growth chamber for 10 days. Water every 3–4 days.

3.2 Plant Spraying

1. Prepare 200 μM MeJA solution and mock solution (as detailed in Subheading 2.3) and transfer each to a separate plastic sprayer.
2. Spray half of the plants with the mock solution until runoff from the sprayed leaves. Cover the plants with a plastic cover and place in a growth chamber.
3. Spray the remaining plants with 200 μM MeJA solution until runoff from the sprayed leaves (*see* **Note 7**). Cover the plants with a plastic cover and place in a separate growth chamber, away from the mock-treated plants.
4. Sample plants from each mock-treated or MeJA-treated groups, 24 h post treatment, as detailed below (*see* Subheading 3.3).
5. Repeat spraying of the remaining plants at 24 and 48 h time-points post first treatment, as detailed in **steps 1–3**.

3.3 Plant Sampling

1. 24 h after the first spraying collect samples from the control and treated groups. Sample 12 plants from the control and 12 plants from the treated group.
2. Uproot the plants carefully to avoid damaging the root system and wash the roots under streaming tap water for 30–60 s to remove the soil (*see* **Note 8**).
3. Cut the root system from the shoot using a scalpel.
4. Collect shoot and root samples for each biological replicate in separate, labelled, 50 mL tubes.
5. Freeze dry the tubes in liquid nitrogen and store at −80 °C until grinding for RNA extraction and purification.
6. Repeat the sampling procedure (**steps 2–5**) for the remaining plants, 48 and 72 h post first spraying treatment.

3.4 RNA Extraction and Purification

1. Using a mortar and pestle, grind plant tissue to a fine powder in liquid nitrogen.
2. After grinding, transfer 100–150 mg of the powder to a microfuge tube, add 1 mL of TRIzol reagent, and vortex for several seconds.
3. Incubate at RT for 5 min.
4. Add 200 μL chloroform to the mixture and vortex for several seconds.

5. Incubate the samples for 10 min at RT to allow the complete dissociation of nucleoprotein complexes.
6. Centrifuge the samples at 14,000× *g* for 20 min at 2–8 °C in a microfuge.
7. Transfer the supernatant to a fresh RNase-free tube.
8. Precipitate the RNA by adding 0.5 mL ice-cold isopropyl alcohol and mixing.
9. Incubate the samples at RT for 10 min, and centrifuge at 14,000× *g* for 10 min at 2–8 °C in a microfuge.
10. Remove the supernatant and wash the RNA pellet by adding at least 0.5 mL ice-cold 80% ethanol. Mix the samples by vortexing, and centrifuge at no more than 5500× *g* for 5 min at 2–8 °C in a microfuge.
11. Repeat **step 9** once more.
12. Briefly dry the RNA pellet (air-dry or vacuum dry for 5–10 min only).
13. Dissolve RNA in RNase-free water by pipetting.
14. Check RNA concentration and quality using NanoDrop and gel electrophoresis (*see* **Note 9**).
15. Store the RNA at −80 °C. RNA samples will be sent out to a service provider for library construction and sequencing.

3.5 cDNA Synthesis for Quantitative Real-Time PCR

1. From the extracted RNA take 2 μg total RNA for DNase treatment and perform the reaction according to the manufacturer's protocol.
2. Prepare the cDNA synthesis reaction mix with the RNA (following DNase) and RNA inhibitor according to the manufacturer's protocol (total volume 20 μL).
3. Load the reaction to a thermal cycler using the following conditions: 25 °C for 5 min (annealing), 42 °C for 60 min (synthesis), 70 °C for 7 min (heat inactivation), 4 °C hold.
4. Prepare a stock cDNA solution by adding 60 μL nuclease-free water to the cDNA, to a final volume of 80 μL.
5. For qRT-PCR reactions, further dilute the stock cDNA solution 1:4.
6. Prepare a premix for each primer pair (test gene 1, test gene 2, housekeeping gene), at a concentration of 5 μM each primer. The premix can be kept in −20 °C for future runs (*see* **Note 10**).
7. Prepare a reaction mix containing: 5 μL SYBR Green (2× concentration mix), 1 μL primer premix, 2 μL nuclease-free water.
8. Add 2 μL of 1:4 diluted cDNA at the bottom of each well.

9. Add 8 μL of the reaction mix to each well. Seal the plate or tubes and mix gently.
10. Briefly centrifuge the plate or tubes to spin down the contents and to eliminate any air bubbles.
11. Load the reaction to the qRT-PCR instrument and program the settings as follows; 3 min at 95 °C, followed by 40 cycles of 10 s at 95 °C and 30 s at 59 °C. Then dissociation analysis with 0.5 °C ramping from 55 to 95 °C.
12. Analyze the results using relevant software. Results from the qRT-PCR experiment can inform on selection of RNA samples to be subsequently used for RNA sequencing.

3.6 Genome-Guided Assembly of RNA-Seq Data and Integration with Gene Clustering Information

1. Using a Linux-based environment, install "miniconda," create a Conda environment, and install the FastQC, Trim Galore, STAR, and Subread tools (*see* **Note 11**).
2. For the RNA-seq.FastQC output files, perform quality control using the FastQC tool [8]. Examine the output through the HTML links to assess data quality and whether additional processing is necessary (*see* **Note 12**).
3. Using the Trim Galore tool [9], trim the sequences.
4. Perform additional quality control on the trimmed files using the FastQC tool.
5. Obtain sequence files (.FASTA) and respective annotation files (.GFF) for the genome assembly version to which the reads will be mapped, from a relevant database (e.g., NCBI, Phytozome, Ensembl Plants) (*see* **Note 13**).
6. Index the FASTA assembly files using the STAR tool [10] (*see* **Note 14**).
7. Align the trimmed RNA-seq reads to the indexed genome, using the STAR tool (*see* **Note 15**).
8. Sort the BAM files that were generated in **step 7** using the SAMtools suite [11].
9. Map the reads to the different genes (feature count) as annotated in the previously acquired GFF files using SAMtools (*see* **Note 16**). Following this step, a reads per gene per sample table will be generated.
10. Normalize the reads count to generate a Trimmed Mean of M-values (TMM) file. Perform this normalization in R, using the DESeq2 tool [12]. This will also require loading the dplyr, ggplot2, readr, openxlsx, and readxl libraries (*see* **Note 17**).
11. Use the plantiSMASH tool, through a standalone version or via the webserver (https://plantismash.bioinformatics.nl/), according to the provided documentation (https://github.com/plantismash), to generate a list of candidate biosynthetic

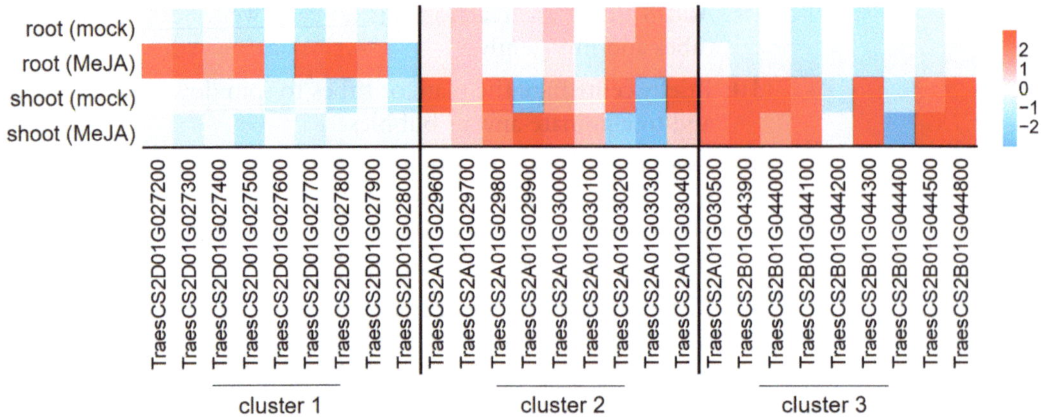

Fig. 2 Integration of gene clustering and gene expression information. Aligning transcriptomic data from MeJA-treated plants with algorithm-defined biosynthetic gene cluster candidates (e.g., by Plantismash) can reveal BGC candidates in which genes are mostly co-expressed and are induced by MeJA (cluster 1), are co-expressed but not induced (cluster 3), or are generally not co-expressed (cluster 2)

gene clusters and gene accessions they are comprised of. Export the list to a text file.

12. Data from the DESeq2 output TMM table and plantiSMASH output text file can be combined in spreadsheet software (e.g., Office Excel), and manually browsed to identify BGCs that contain co-expressed and/or MeJA-induced genes (Fig. 2).

4 Notes

1. Distilled water (DW) is sufficient for the surface sterilization reaction.
2. Since the MeJA solution is not soluble in water it must be first diluted in absolute ethanol. After the MeJA ethanolic solution is first prepared, slowly add it dropwise to the water while continuously stirring.
3. A fine mesh strainer can be used to avoid losing seeds during the washes.
4. Use tweezers to spread the seeds apart on the wet filter paper. Several plates can be used.
5. The plates can be incubated at 4 °C for up to 4 nights, for cold stratification of the seeds.
6. During the incubation of the Petri dish in the growth chamber, the paper tends to dry out. Add 1–2 mL water when necessary and reseal the plates.
7. MeJA is highly volatile. It is therefore advised to keep mock-treated plants away from MeJA-treated plants at all times.

8. The roots are washed from the soil under a gentle stream of water to ensure the intactness of the roots. The water temperature should be no more than room temperature.
9. The expected 260/280 and 260/230 nm values in the Nanodrop readings are typically in the range of 1.8–2.0 and 2.0–2.2 (respectively). If the ratio is substantially lower than expected, it may indicate the presence of contaminants. In this case an ethanolic precipitation protocol needs to be performed for obtaining high-purity RNA.
10. To achieve accurate and reproducible expression profiling of selected genes using real-time PCR, an internal control gene (housekeeping gene) is required for expression normalization, which exhibits stable expression and is not affected by MeJA treatment. Here we use tubulin as the housekeeping gene.
11. When creating Conda environments, a single environment can be opened to contain the four tools, or individual environments for each tool. We chose the latter option.
12. Out of the many parameters examined, those that are important and require further attention are the "basic statistics," where the sequencing depth can be seen in the "total sequencing" tab, "per base sequence quality," and "overrepresented sequences."
13. In the Ensembl Plants website, there are three types of genomes: Unmasked, soft-masked, and hard-masked. For our analysis, we found that the unmasked version gave the highest mapping percentage.
14. This step requires a large allocation of RAM. We used a remote server, allocating 150 GB memory.
15. At the end of this step, assess the percentage of reads that aligned to the genome. Our samples had between 72% and 87.5% uniquely mapped reads, with the majority mapping above 80%. However, above 65% is acceptable and the mapping percentage may vary due to biological parameters.
16. Following this step, the percentage of mapped reads can be examined. An additional reduction of the uniquely mapped reads' percentage is expected, due to reads mapping to unannotated regions and overlapping annotations.
17. In addition to the TMM normalization, which normalizes to sequencing depth, additional normalization to gene length may be performed for FPKM normalization.

Acknowledgments

The GP Lab is funded by the Israel Science Foundation (315/24). BN is funded by a Shamir fellowship, Israeli Ministry of Innovation, Science and Technology (0007045).

References

1. Polturak G, Liu ZH, Osbourn A (2022) New and emerging concepts in the evolution and function of plant biosynthetic gene clusters. Curr Opin Green Sustain Chem 33. https://doi.org/10.1016/j.cogsc.2021.100568
2. Kautsar SA et al (2017) plantiSMASH: automated identification, annotation and expression analysis of plant biosynthetic gene clusters. Nucleic Acids Res 45(W1):W55–W63. https://doi.org/10.1093/nar/gkx305
3. Topfer N, Fuchs LM, Aharoni A (2017) The PhytoClust tool for metabolic gene clusters discovery in plant genomes. Nucleic Acids Res 45(12):7049–7063. https://doi.org/10.1093/nar/gkx404
4. Schlapfer P et al (2017) Genome-wide prediction of metabolic enzymes, pathways, and gene clusters in plants. Plant Physiol 173(4):2041–2059. https://doi.org/10.1104/pp.16.01942
5. Yamane H (2013) Biosynthesis of phytoalexins and regulatory mechanisms of it in rice. Biosci Biotechnol Biochem 77(6):1141–1148. https://doi.org/10.1271/bbb.130109
6. Liu Y et al (2024) Hordedane diterpenoid phytoalexins restrict Fusarium graminearum infection but enhance Bipolaris sorokiniana colonization of barley roots. Mol Plant 17(8):1307–1327. https://doi.org/10.1016/j.molp.2024.07.006
7. Polturak G et al (2022) Pathogen-induced biosynthetic pathways encode defense-related molecules in bread wheat. Proc Natl Acad Sci USA 119(16):e2123299119. https://doi.org/10.1073/pnas.2123299119
8. Andrews S (2010) FastQC: a quality control tool for high throughput sequence data. Available online at: http://www.bioinformatics.babraham.ac.uk/projects/fastqc
9. Krueger F (2015) Trim galore. A wrapper tool around Cutadapt and FastQC to consistently apply quality and adapter trimming to FastQ files. 516(517):517
10. Dobin A et al (2013) STAR: ultrafast universal RNA-seq aligner. Bioinformatics 29(1):15–21. https://doi.org/10.1093/bioinformatics/bts635
11. Li H et al (2009) The sequence alignment/map format and SAMtools. Bioinformatics 25(16):2078–2079. https://doi.org/10.1093/bioinformatics/btp352
12. Love M, Huber W, Anders S (2014) Moderated estimation of fold change and dispersion for RNA-seq data with DESeq2. Genome Biol 15. https://doi.org/10.1186/s13059-014-0550-8

Chapter 5

Investigating Intraspecific Diversity of Metabolic Gene Clusters with MLPA and Synthetic Probes

Viktoriia Plyhun, Anastasiia Satyr, and Agnieszka Zmienko

Abstract

Multiplex Ligation-Dependent Probe Amplification (MLPA) is a robust method for efficient genotyping known structural variations, especially duplications and deletions, in a population scale. It allows for the simultaneous assessment of multiple regions in the genome in a single assay. This feature makes MLPA especially useful in the analysis of metabolic gene clusters (MGCs), where three or more neighboring genes need to be genotyped to decipher the overall MGC structural pattern. By enabling fast and cost-effective screening of a large number of individuals or lines, MLPA may measure MGC intraspecific variation or reveal new MGC member(s), e.g., resulting from gene duplication. This in turn may provide insight into the potential diversity of biosynthetic pathways encoded by a given MGC. Currently, however, custom synthetic probes are required to target non-human genes with MLPA. Here we describe the process of designing such synthetic probes detecting MGC genes and conducting MLPA assays. MLPA with synthetic probes may be applied for MGC analysis in various species including plants, animals, or fungi.

Key words MLPA, MGC, Synthetic probes, Duplication, Deletion, Population analysis, Structural variation

1 Introduction

Metabolic gene clusters (MGCs) consist of three or more neighboring genes involved in a common metabolic pathway. Recent years marked a significant progress in the MGC discovery, especially in the genomes of plant and fungal species [1, 2]. Metabolic compounds biosynthesized by the enzymatic pathways encoded in the MGCs may play a role in the developmental processes as well as in the response to various biotic and environmental factors [3–8]. Increasing evidence indicates that MGCs may display substantial level of intraspecific diversity [9]. Among them, copy number variations, including gene duplications and deletions, are important contributors to this diversity [10, 11]. Deletion of an MGC gene member may lead to alteration in the metabolic products biosynthesized by this MGC. Similarly, duplicated genes may be

Hans-Wilhelm Nützmann and Ramesha Thimmappa (eds.), *Metabolic Gene Clusters: Methods and Protocols*, Methods in Molecular Biology, vol. 3054, https://doi.org/10.1007/978-1-0716-5384-5_5,

slightly divergent in sequence and biochemical function from their parental homologues, which might lead to metabolic diversity. Gene duplication and subsequent neo-functionalization are considered to be the main drivers in generating metabolic branchpoints [12]. Therefore, screening large populations for variation in MGC genes may promote the identification of new metabolic transformations and increase our understanding of the processes driving specialized metabolite evolution [11, 13, 14].

Multiplex Ligation-Dependent Probe Amplification (MLPA) is a molecular biology method that identifies structural variations in DNA, primarily duplications and deletions. It has been introduced in 2002 [15] and subsequently developed by MRC Holland company. MLPA detects the presence or lack of the specific sequence targeted by MLPA probes. Due to their specific design, many MLPA probes with unique lengths may be created and combined into a larger set (a probemix). Commercial SALSA® MLPA® Probemixes consist of up to 60 MLPA probes (https://www.mrcholland.com/). This multiplexing ability makes MLPA a perfect choice in situations when several genes or regions need to be genotyped to get the overall picture of the structural diversity at the genomic loci related to a common condition or process. Accordingly, MLPA is well-suited for simultaneous genetic profiling of MGC gene members and detecting variants in MGC composition or structure [11]. However, currently available commercial MLPA probemixes detect copy number changes and genetic disorders in human genomes only, while for other species custom probes need to be created. We successfully use home-designed short synthetic probes (hereafter: Probes) for that purpose [16–18].

In a typical experiment, we use up to 31 Probes, including Probes specific to the regions of interest and the Control Probes. The Probes have fixed length between 90 and 200 nt, with 3-base length increments for Probes up to 120 nt and 4-base increments for longer Probes (Table 1). Each Probe is composed of two short oligonucleotides (Half-Probes), left and right, which also have fixed lengths. Each Half-Probe contains a target-specific sequence (TSS), a primer-annealing site, and an optional stuffer sequence, joined in the following order (Fig. 1a):

5′-left primer-annealing site—left stuffer—left TSS-3′ (left Half-Probe)

and

[Phospho]-5′-right TSS—right stuffer—right primer-annealing site-3′ (right Half-Probe).

Primer-annealing sites bind the respective primers (one of them is fluorescently labeled) present in the SALSA® MLPA® Reagent Kit. The stuffer sequence is inserted to ensure that the length of

Table 1
List of the recommended synthetic probes' lengths for the MLPA assay

Probe length [nt]			Left half-probe element length [nt]			Right half-probe element length [nt]		
Total	Left half-probe	Right half-probe	Left primer-annealing site	Left stuffer	Left TSS	Right TSS	Right stuffer	Right primer-annealing site
90	45	45	19	0–5	21–26	21–22	0–1	23
93	46	47	19	0–6	21–27	21–24	0–3	23
96	48	48	19	0–8	21–29	21–25	0–4	23
99	49	50	19	0–9	21–30	21–27	0–6	23
102	51	51	19	0–11	21–32	21–28	0–7	23
105	52	53	19	0–12	21–33	21–30	0–9	23
108	54	54	19	0–14	21–35	21–31	0–10	23
111	55	56	19	0–15	21–36	21–33	0–12	23
114	57	57	19	0–17	21–38	21–34	0–13	23
117	58	59	19	0–18	21–39	21–36	0–15	23
120	60	60	19	0–20	21–41	21–37	0–16	23
124	62	62	19	0–22	21–43	21–39	0–18	23
128	64	64	19	0–24	21–45	21–41	0–20	23
132	66	66	19	0–26	21–47	21–43	0–22	23
136	68	68	19	0–28	21–49	21–45	0–24	23
140	70	70	19	0–30	21–51	21–47	0–26	23
144	72	72	19	0–32	21–53	21–49	0–28	23
148	74	74	19	0–34	21–55	21–51	0–30	23
152	76	76	19	0–36	21–57	21–53	0–32	23
156	78	78	19	0–38	21–59	21–55	0–34	23
160	80	80	19	0–40	21–61	21–57	0–36	23
164	82	82	19	0–42	21–63	21–59	0–38	23
168	84	84	19	0–44	21–65	21–61	0–40	23
172	86	86	19	0–46	21–67	21–63	0–42	23
176	88	88	19	0–48	21–69	21–65	0–44	23
180	90	90	19	0–50	21–71	21–67	0–46	23
184	92	92	19	0–52	21–73	21–69	0–48	23
188	94	94	19	0–54	21–75	21–71	0–50	23
192	96	96	19	0–56	21–77	21–73	0–52	23
196	98	98	19	0–58	21–79	21–75	0–54	23
200	100	100	19	0–60	21–81	21–77	0–56	23

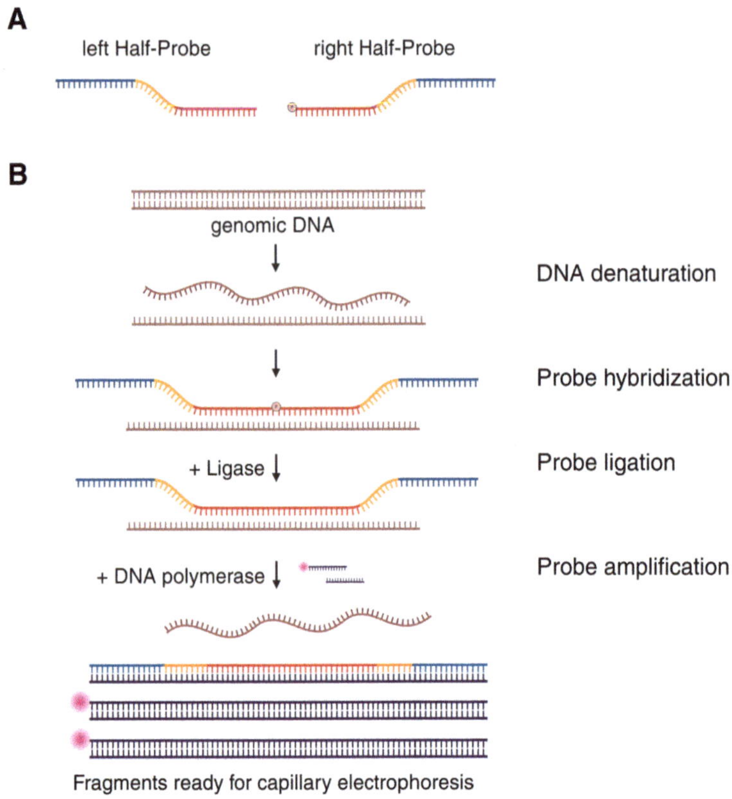

Fig. 1 Overview of MLPA experiment. (**a**) Composition of the synthetic left and right Half-Probes. Each Half Probe constitutes of target-specific sequence (red), optional stuffer (yellow), and primer annealing sequence (blue). Additionally, 5′ end of the right Half-Probe is phosphorylated. (**b**) Stages of the MLPA assay. See main text for the description

each Probe in the assay will be unique. TSSs of the corresponding left and right Half-Probes are complementary to the directly adjacent regions in the genomic DNA. The right Half-Probe has its 5′-end phosphorylated to enable ligation with the 3′-OH end of the left Half-Probe, following their proper hybridization to the genomic DNA.

Selecting TSSs is the crucial part of the entire Probe design process and determines their specificity and sensitivity in detecting the exact variant of interest (e.g., gene deletion, specific polymorphic site in the sequence, etc.). The exact position of the TSS within the target gene also depends on the Probe type. MLPA is a relative technique, where the quantity of the investigated DNA region in the sample is determined by comparison to the reference sample (s) with known genotype, assayed with the same set of Probes. To enable sample-to-sample comparison, Control Probes must be included in each assay. TSSs in Control Probes should target regions free of any sequence polymorphisms or copy number variations and detect genes that are unrelated to the investigated MGC (preferably located on a different chromosome). On the contrary, Probes detecting duplications or deletions should contain TSSs that overlap with the region of investigated variation. In any case, however, the corresponding left and right Half-Probes should be designed simultaneously, starting from the determination of the TSS elements. Following design and synthesis, multiple Probes may be combined in a custom Probe Set Mix and used along with SALSA® MLPA® Reagent Kits (MRC Holland). The two-day assay is a multistep procedure and follows MLPA General Protocol (Fig. 1b):

Probe Hybridization to Denatured Genomic DNA A sample of denatured genomic DNA is incubated with the Probe Set Mix for at least 16 h. During this step, the TSS parts of the Half-Probes hybridize to the complementary regions in the genomic DNA. The amount of the hybridized oligonucleotides depends on the amount of the genomic DNA in the solution, its ploidy, and the number of the target sites in the genome.

Probe Ligation Half-Probes that hybridized to genomic DNA are ligated by adding DNA Ligase to the sample and short incubation, which produces a full-length Probe. Next, the enzyme is heat-inactivated. Only properly positioned oligonucleotides, with 3′-OH end of the left Half-Probe located immediately adjacent to 5′-phosphorylated end of the right Half-Probe may be connected by the enzyme. Any mismatches around the ligation site will also prevent or negatively affect the ligation efficiency of the adjacent Half-Probes. Combined, the hybridization and ligation steps ensure high specificity of the MLPA assay.

Probe Amplification All Probes reconstituted after successful ligation are amplified in a single PCR reaction, using a pair of primers complementary to the primer-annealing sites present in each Probe. The resulting amplicons are fluorescently labeled with FAM or Cy-5, depending on the SALSA® MLPA® Reagent Kit used. A common PCR step enables Probe multiplexing.

Following amplification, the fragments are separated with a single-nucleotide resolution by capillary electrophoresis in the appropriate polymer matrix and the fluorescence signal is collected. Size standard is added to each sample to allow subsequent identification of the fragments' lengths. Each Probe in the assay gives rise to fragments of unique size, specified by the assay design. Therefore, it is possible to simultaneously determine the presence/absence/copy number of multiple target regions in many individuals, based on fragment peak analysis and comparison among the samples. As a result, structural pattern of the entire MGC may be reconstituted in a single assay. We successfully used this approach to screen a large *A. thaliana* population for copy number variations in the thalianol, marneral, and arabidiol/baruol MGCs [11]. It should be noted that MLPA is designated to identify known variant types in the previously uncharacterized samples. It is of limited use in detecting new variations and polymorphisms. Nevertheless, observing unusual signals from MLPA probes may provide important clues about such new variants as well. For example, we were able to link some "noisy" MLPA results to the presence of previously undetected gene duplication in *A. thaliana* arabidiol/baruol metabolic gene cluster, thus demonstrating the power of MLPA approach in MGC analysis [11].

2 Materials

2.1 MGC Sequence Information

1. A list of genes and the sequence of your MGC of interest, usually in the reference genome for a given species.
2. Variant coordinates and the alternative allele sequence for the polymorphisms you want to detect within MGC gene members.

2.2 DNA Samples

Genomic DNA samples from the individuals/strains/lines/accessions, etc., for which MGC variation pattern is to be determined. DNA should be dissolved in nuclease-free water or 10 mM Tris-Cl, pH 8.5. DNA concentration should be identical for all samples (typically 0.4–50 ng/μL, which will result in 2–250 ng DNA per sample per assay) [17, 19, 20].

2.3 Synthetic Oligonucleotides

Custom synthetic oligonucleotides (Half-Probes) targeting the genes of interest. They should be designed according to the description in the Methods section and purchased from the external provider.

2.4 MLPA Assay

1. SALSA® MLPA® Reagent Kit (MRC Holland).
2. Nuclease-free water.
3. 10 mM Tris-Cl, pH 8.5 (optionally).
4. 0.2 mL 8-strip PCR tubes (recommended) or, alternatively, standard 0.2 mL PCR tubes (*see* **Note 1**).
5. Extra 8-strip PCR caps.
6. Thermocycler with heated lid.
7. Benchtop minicentrifuge, with adapters for PCR strips.
8. Vortex mixer.
9. Standard and multichannel pipettes.

2.5 Data Collection and Analysis

1. Access to the capillary electrophoresis instrument, e.g., Genetic Analyzer (Applied Biosystems) or similar, usually through a core facility or a commercial service.
2. Data analysis software for fragment analysis with the MLPA analysis option, e.g., GeneMarker® (SoftGenetics) or Coffalyser.Net™ (MRC Holland).

3 Methods

3.1 Probe Design

3.1.1 Selecting TSSs

1. Get the sequence of your target gene, including the information about the intron-exon boundaries, e.g., by retrieving it from the reference genome record in the respective database, like GenBank.
2. Use *blast* to locate this gene in other available genomes of the same species and extract the respective sequence. If you aim to detect a specific sequence variant, e.g., SNP, include the exemplar genomic sequences for both the reference and alternative allele.
3. Perform multiple sequence alignment, e.g., using online EMBL-EBI toolkit (https://www.ebi.ac.uk/jdispatcher/msa). Find polymorphic positions (SNPs, indels, etc.) located within the gene exons and mark them on the reference sequence, e.g., in the Word document.

   Depending on the species of interest, simpler approaches may be applied to detect and mark the polymorphisms in the analyzed sequence (*see* **Note 2**).
4. Inspect the results and select the candidate region as follows (Fig. 2):
   (a) If you design a Control Probe or aim to detect a whole-gene deletion, simply select an exon or its part (~200 nt) that has no SNPs and other sequence variations in the alignment.
   (b) If you aim to detect a partial gene deletion, the selected region must additionally fully overlap with the deletion.
   (c) If you aim to detect whole-gene or partial gene duplications where all copies are known to have identical sequence, follow the same instructions as for the deletion variants.
   (d) If you want to distinguish two closely related but not identical sequences, e.g., a gene and a pseudogene or a gene and its divergent duplicate copy, center the candidate sequence on the polymorphic region and extend it in both directions, to obtain at least 50 nt sequence. Then proceed to selecting the left and right TSSs, as described

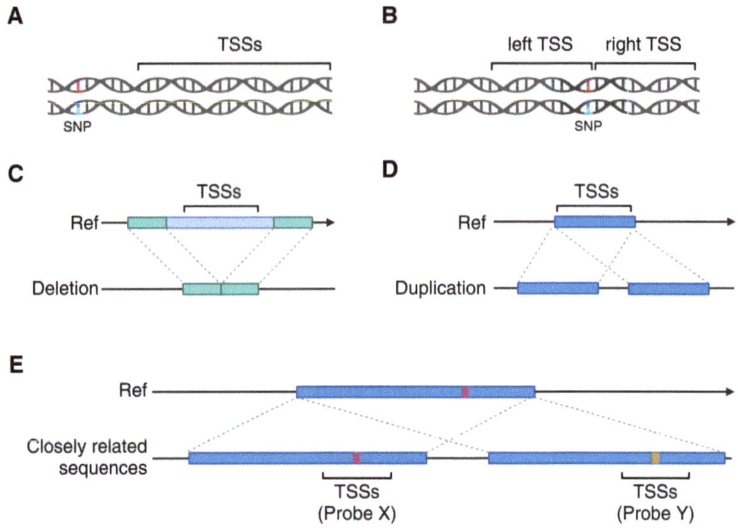

Fig. 2 Examples of locating TSSs in the genomic DNA for different probe types. (**a**) TSSs in conserved regions (Control Probes); (**b**) the 3′ end of the left TSS spans the polymorphic position (SNP detection); (**c**, **d**) TSSs overlap deletion/duplication region (copy number variation detection); (**e**) two Probes with TSSs targeting the same region but different allele (distinguishing between the alternative sequence variants)

below, but make sure to keep the divergent sequence around the ligation site. If the reference and the alternative alleles differ by one nucleotide only, make sure this polymorphism is located exactly at the 3′ end of the left TSS (*see* **Note 3**).

5. Within the chosen region, select two sequence strings, each 21 nt or longer, as your left and right TSS candidates. Adjust their positions and lengths according to the following rules:
   (a) The left and the right TSSs must be directly adjacent to each other. This means that when you move the 3′end of your left TSS candidate, you must move the 5′ end of your right TSS candidate accordingly.
   (b) The melting temperature (Tm) of your candidate TSSs should be close to 70–71 °C and not lower than 67 °C, when calculated with the OligoAnalyzer Tool (https://eu.idtdna.com/calc/analyzer), using the following parameters: Oligo Conc 1 μM, Na+ Conc 100 mM, Mg++ Conc 3 mM, dNTP Conc 0.8 mM. Adjust TSSs lengths or reposition them if needed.
   (c) Avoid including long homopolymer tracts and GC tracts $\geq$4 nt in the TSSs. Reposition TSSs if needed (*see* **Note 4**).
6. After selecting the left and right TSS candidates, join their sequences. The combined sequence should be between

42 and 158 nt. Use it as a query in *blastn* search against the respective reference genome. A well-designed pair of TSSs should perfectly match only the desired region and not any other sequence in the genome. The presence of the non-target hits may be sometimes acceptable as long as they have 4 or more mismatches and/or gaps in the sequence alignment around the ligation area. If the *blastn* results will indicate low specificity of the TSSs, repeat **step 5**, or chose another exon.

7. Repeat the above procedure for every gene of interest and every region you want to target.

3.1.2 Creating Half-Probes

1. Select the recommended Probe length from the list in Table 1. Make sure that no other Probe in the current assay has the same length already and that the lengths of TSSs are within the limits defined for that Probe (*see* **Note 5**).
2. Add stuffer sequence to each TSS, if necessary. Stuffer is created by inserting the desired amount of bases from the M13 phage sequence (GenBank ID V00604, range 3:119), which has no similarity to known eukaryotic DNA:

   CGCTACTACTATTAGTAGAATTGATGCCACCTTTT CAGCTCGCGCCCCAAATGAAAATATAGCTAAACAGGT TATTGACCATTTGCGAAATGTATCTAATGGTCAAAC TAAATCTAC

   For each Half-Probe, the length of the stuffer is determined by subtracting the length of the TSS and primer-annealing sequence from the anticipated Half-Probe length:

   left stuffer = [left Half-Probe] − [left TSS] − 19, where 19 is the length of the left primer-annealing sequence

   right stuffer = [right Half-Probe] − [right TSS] − 23, where 23 is the length of the right primer-annealing sequence

   Left stuffer sequence should be added to 5′ end of the left TSS. Right stuffer sequence should be added to the 3′ end of the right TSS (Fig. 1a).
3. Add primer-annealing sites to the respective TSSs.

   The sequence of the left primer-annealing site is:
   GGGTTCCCTAAGGGTTGGA
   It should be added to 5′ end of the left stuffer.
   The sequence of the right primer-annealing site is:
   TCTAGATTGGATCTTGCTGGCGC
   It should be added to the 3′ end of the right stuffer.

3.1.3 Oligonucleotide Synthesis

1. Once the Half-Probe sequences are complete, order the oligonucleotide synthesis from the external service, at a small scale. All oligos must be purified by HPLC or PAGE. Additionally, the 5′ end of each right Half-Probe oligo must be phosphorylated.

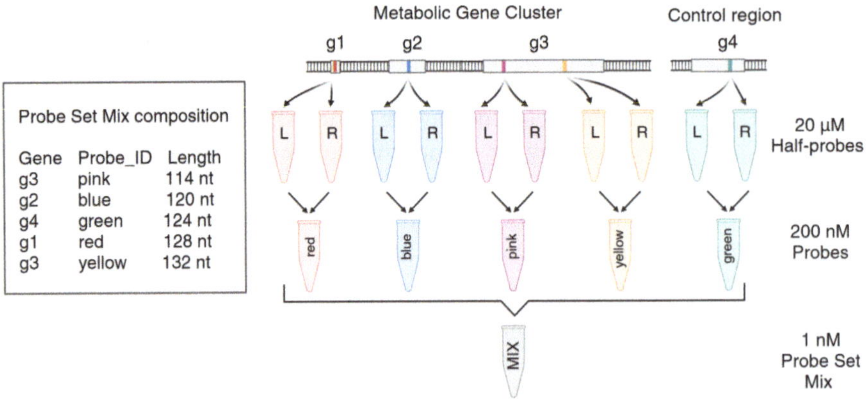

Fig. 3 Preparation of a probe set mix for an MLPA assay

2. Upon arrival, prepare individual 20 μM Half-Probe oligo stocks by adding nuclease-free water or 10 mM Tris-Cl, pH 8.5. Store them at −20 °C.

3.2 Preparing the Probe Set Mix

(If your Probe Set Mix has been already prepared, skip this section and proceed directly to Subheading 3.3).

A Probe Set Mix is a composition of all synthetic Half-Probes included in the single assay. For each gene, at least one pair of Half-Probes must be included. We also recommend including three to five Control Probes. The concentration of each Half-Probe in the final Probe Set Mix is 1 nM, which requires 20,000 × dilution of the 20 μM stocks (Fig. 3).

1. Thaw all 20 μM Half-Probe stocks intended to constitute one assay, as specified in the assay Probe list (*see* **Note 5**).
2. For each Probe separately, prepare 200 nM solution of the respective oligonucleotide pair in one Eppendorf tube, by adding add 2 μL of the left Half-Probe and 2 μL of the right Half-Probe to 196 μL nuclease-free water.
3. In a single Eppendorf tube, combine 2 μL of each 200 nM Probe from the previous step and fill up to 400 μL with nuclease-free water. The resulting Probe Set Mix contains all desired Half-Probe pairs, and the concentration of each Half-Probe is 1 nM. For example, if you prepare a Probe Set Mix for three control genes and five MGC genes, you need to mix 8 × 2 μL of 200 nM Probe solutions = 16 μL. You must add 384 μL water to obtain the final Probe Set Mix. The 400 μL volume is sufficient for assaying ~250 samples (*see* **Note 6**).

3.3 Performing MLPA Assay

All incubations in this multi-step assay are performed in a pre-programmed thermocycler. The maximal number of samples that can be simultaneously assayed depends on the thermocycler capacity (typically 96) and on the experimenter's skills. Sample treatments (e.g., enzyme addition) are required between the steps.

Table 2
Thermocycler settings for MLPA assay

MLPA assay stage	Program step	Thermocycler settings
Denaturation	Step 1	98 °C, 5 min; Ramp 2 °C/s
	Step 2	25 °C, ∞
Hybridization	Step 3	95 °C, 1 min; Ramp 2 °C/s
	Step 4	60 °C, 16–20 h
	Step 5	54 °C, ∞
Ligation	Step 6	54 °C, 15 min
	Step 7	98 °C, 5 min
	Step 8	20 °C, ∞
Amplification	Step 9	95 °C, 30 s; Ramp 2 °C/s
	Step 10	60 °C, 30 s
	Step 11	72 °C, 1 min; Ramp 2 °C/s
	Step 12	GOTO Step 9, 34×
	Step 13	72 °C, 20 min
	Step 14	4 °C, ∞

3.3.1 Day One (Hands-on Time 30–90 min, Depending on the Number of Samples)

1. Create the MLPA program in the thermocycler according to the instructions in Table 2. This program will include all incubation steps for the two-day assay.
2. Pre-heat the thermocycler. Use heated lid option.
3. Thaw and spin down MLPA buffer from the SALSA MLPA Reagent Kit and a vial of the appropriate Probe Set Mix, prepared according to the instructions in Subheading 3.2. Keep them on ice.
4. Thaw and spin down genomic DNA samples that will be assayed. Keep the same DNA concentration for all future samples added to the project (*see* **Note** 7).
5. Create a sample list (a paper or electronic one) and assign a consecutive number to each DNA sample. Label the 8-tube PCR strips with the corresponding numbers. It is sufficient to label only the first and the last tube in each strip (e.g., 1 and 8 for the first strip, 9 and 16 for the second strip, etc.). Arrange the strips on the 96-tube rack and keep them at room temperature while setting up the assay.
6. For each sample, add 5 μL of genomic DNA to the respective strip tube. Close the strips with the new caps (*see* **Note 8**).
7. Insert the tubes into the thermocycler and run MLPA program. Heat the samples to 98 °C for 5 min then cool them to 25 °C (Program Steps 1 and 2). This is the DNA Denaturation Stage.

8. Remove the strips from the thermocycler and briefly spin down. Keep them at room temperature and immediately proceed to the next step.
9. Prepare Master Mix I. In an Eppendorf tube, combine 1.5 μL MLPA buffer with 1.5 μL Probe Set Mix per sample. Prepare enough mix for all samples, with 10% surplus. Briefly vortex and spin down the tube.
10. Add 3 μL Master Mix I to each sample. Close the strips with the new caps, briefly mix by inverting the strips up and down or vortexing, then spin down.
11. Insert the strips into the thermocycler and resume MLPA program by skipping to the next step. Heat the samples to 95 °C for 1 min and then incubate at 60 °C for 16–20 h (Program Steps 3 and 4). This is the Probe Hybridization Stage.

3.3.2 Day 2 (Hands-on Time 20–40 min, Depending on the Number of Samples)

1. After hybridization, let the samples cool down to 54 °C (Program Step 5). CAUTION: Do not remove the samples from the thermocycler, as this will affect probe hybridization efficiency and specificity.
2. Thaw and spin down Ligase Buffer A, Ligase Buffer B, and SALSA PCR Primer Mix. Keep them on ice.
3. Prepare Master Mix II. In an Eppendorf tube, combine 3 μL Ligase Buffer A, 3 μL Ligase Buffer B, and 25 μL nuclease-free water per sample. Vortex and spin down, then add 1 μL SALSA Ligase-65 per sample. Mix by pipetting (do not vortex Master Mix II after the enzyme has been added) and spin down. Prepare enough mix for all samples, with 10% surplus.
4. Open the thermocycler lid and add 32 μL of Master Mix II to each sample, without removing them from the thermocycler. Mix gently by pipetting up and down three times. We advise to open only one strip at a time and close it with the new cap, right after adding Master Mix II, to reduce sample evaporation.
5. Close the thermocycler and resume the program by skipping to the next step. Continue incubation of the samples at 54 °C for 15 min (Program Step 6). This is the Probe Ligation Stage. Next inactivate the enzyme by heating the samples to 98 °C for 5 min (Program Step 7). Cool the samples to 20 °C (Program Step 8), remove them from the thermocycler, and spin down. Keep the strips at room temperature and directly proceed to the next step.
6. Prepare Master Mix III. In an Eppendorf tube, combine 2 μL SALSA PCR Primer Mix and 7.5 μL nuclease-free water per sample. Vortex and spin-down, then add 0.5 μL SALSA Polymerase per sample. Mix by pipetting (do not vortex the mix

after the enzyme has been added) and spin down. Prepare enough mix for all samples, with 10% surplus.

7. Add 10 μL of Master Mix III to each sample, mix by pipetting up and down several times. Close the strips with the new caps and spin down. At this stage, the volume of the samples should be 50 μL each.
8. Insert the samples into the thermocycler and resume MLPA program by skipping to the next step. Run 35 cycles of PCR: sample denaturation at 95 °C for 30 s, annealing at 60 °C for 30 s, and elongation at 72 °C for 1 min (Program Steps 9–12). This is the Probe Amplification Stage. Continue with final elongation at 72 °C for 20 min (Program Step 13), then cool down the samples to 4 °C (Program Step 14).
9. Remove the tubes from the thermocycler and proceed to fragment separation by capillary electrophoresis (*see* Subheading 3.4). Alternatively, samples may be stored at 4 °C for up to 1 week or at −20 °C for longer periods prior to fragment separation. Protect the samples from light!

3.4 Fragment Analysis by Capillary Electrophoresis

This step is usually performed by the core facility or the external service provider. Details of the sample preparation and fragment separation will therefore depend on the type of the instrument used. Follow the standard recommendations for fragment separation of Cy5 or FAM-labeled products on a given instrument. It may be necessary to optimize run conditions first, for example, adjust the amount of injected sample to obtain the peak intensity that will fall within the detection range. For a given instrument, the optimized run parameters may be then applied for all samples and all MLPA Assays, as long as the species and input DNA amount remain the same, plus no unusually high DNA copy number is expected for any gene. For example, for resolving Arabidopsis FAM-labeled MLPA samples on ABI Prism 3130xl Genetic Analyzer (Applied Biosystems), we use 0.5 or 1 μL sample, which is mixed with 9 or 8.5 μL Hi-Di formamide, respectively, and 0.5 μL GeneScan™ 600 LIZ® size standard. The samples are denatured at 95 °C for 5 min and cooled down to 4 °C. Each sample is loaded in 36-cm capillary with POP7 separation matrix at 1.2 kV (injection) and run at 15 kV (electrophoresis). The results are written into.fsa files.

3.5 Data Analysis

The output files from the fragment electrophoresis instrument need to be processed with the dedicated software. The analysis involves calculating fragment size based on the internal size standard, Probe identification (this step requires information about the Probes in the Probe Set Mix), followed by intrasample normalization of the identified fragment peaks. The last step is inter-sample comparative analysis and variant identification. There are various free or commercial programs capable of the basic analysis of the capillary electrophoresis output files. However, we recommend using software with the MLPA analysis option built-in, which greatly simplifies the analysis, supports adding new samples to the already analyzed set, and improves the reproducibility. Below we briefly describe the main analysis steps of MLPA data using exemplar data generated for *A. thaliana* samples and Probe Set Mix targeting the thalianol biosynthesis gene cluster (Table 3). We demonstrate how to perform the analysis with two programs: GeneMarker® (SoftGenetics) and Coffalyser.Net™ (MRC Holland). In our experience, the analysis results of MLPA data generated using synthetic Probe Set Mix reported by both programs are congruent in terms of fragment size analysis, variant detection, and copy number ratios.

3.5.1 Data Analysis with GeneMarker

The free-trial version of the GeneMarker software is available for download from the SoftGenetics website (https://softgenetics.com/). You will need a license to use the software after the trial period. We present how to perform the main steps of the MLPA data analysis in Gene Marker V.2.4.2. Read the Reference Manual for the full description of the program version you use and its options.

1. Load raw.fsa files into GeneMarker (select *File* → *Open Data* from the menu).
2. Open Run Wizard dialog box (select *Project* → *Run* from the menu). Set *Analysis Type* to "MLPA" and select the appropriate *Size Standard* and *Standard Color* from the built-in list of common internal lane standards. Choose standard that has been added to your samples during the capillary electrophoresis (in our case "GS600").
3. Select the appropriate Panel to inform the software about the expected lengths of the Probes in the assay. Any peak found in the bin assigned to a given Probe will be automatically labeled by its name, as provided in the Panel. *See* **Note 9** on how to proceed if you are going to analyze the samples for a brand new assay, for which the Panel has not been created yet.
4. Select "*Internal Control Probe Normalization*" option. To use this option effectively, you need to mark your Control Probes as Controls in the selected Panel (*see* **Note 10**). The software will use the signal from all Control Probes in the sample to adjust the peak intensities of the remaining Probes to the same scale for comparative purposes.
5. We recommend not to change other parameters of the software if you are a beginner.
6. Run the analysis.

Table 3
Probes targeting thalianol biosynthesis gene cluster in *A. thaliana* used in the exemplar assay

Expected length	Probe name	Gene	Left TSS	Right TSS
93	A010	*AT5G47980*	TTATATGGCGATGTGCCAGAAATGCCT	CACACGCTAACTTGATCGTCCCAA
96	ctrl1	*AT1G01040*	CAAGAATCATACCTCAGTGAGGTGGTGT	CACTCTTGCAATGTGAGCTTCTGGA
99	C030	*AT5G48000*	AGTCGCATGTCTCTCCGATGATCG	GTAAACCCATTGATCCCGGTGGTAA
105	E111	*AT5G48010*	GCTCCATTTTCCGACCTATAGGAACAC	GTTCTCTCTTTGGGCTCTCACACATTAC
108	E140	*AT5G47950*	TGATTTGATCTCGCAAGAACAAACGTCTCATAC	TCTCAAAACTTCCCTGTCCGAGATCTTGAC
111	ctrl2	*AT4G21580*	CTTTCCCTGAAGTGGCATGCACT	GTTTGGTCTACTGTCTTCATGATGGGC
114	E141	*AT5G47950*	AGATTGTTTGGTGAAATCATTAATTCCAAGTTTAACC	GCGAAGATCATGACATGCATTCTGTAACTAGC
120	E142	*AT5G47950*	AGATTGTTTGGTGAAATCATTAATTCCAAGTTTAACT	TCGAGCTTCATGACATGCATTCTGTAACTAGC
124	ctrl3	*AT2G36230*	CTTTCCTTCAACGTCTACGCCATG	CACCAGAAACTCATCTGAAAATCCTCC
132	E091	*AT5G47990*	AAGATTGCATCCTCCGGGACCA	GTGGTAAGGACGTTCAAAGAGACATGTG
144	ctrl4	*AT5G23290*	TGGCAGATGAAGCTGGAATGGTCTT	GCAGGCCAAAGTTAAACAGTTAACAGCTG
148	E110	*AT5G48010*	TGGACCGAAGGCTGGAGAGGATACT	CACCTGTTCACCACCAACAACTATGC
160	E090	*AT5G47990*	TTATACTCTTTGTCCCAATCCACCAGTCTTTC	AAGAACATCTCATCTAAATATGGACCTCTCCTCC
164	A011	*AT5G47980*	AACAAGGCTGAGAAGCTCTCTATCCCAG	GCGCTGTCACGTTTCTATCCTCTC
172	ctrl5	*AT1G73010*	GGTTCGAGTTGATCTCAGAGAAGAGTTC	GCTAATTCCGAGATGTTCAACGATGGT
188	C031	*AT5G48000*	CGACACACTGTCTCACTCCACCT	CCAAACACCATGAACGTCTTAGATCC

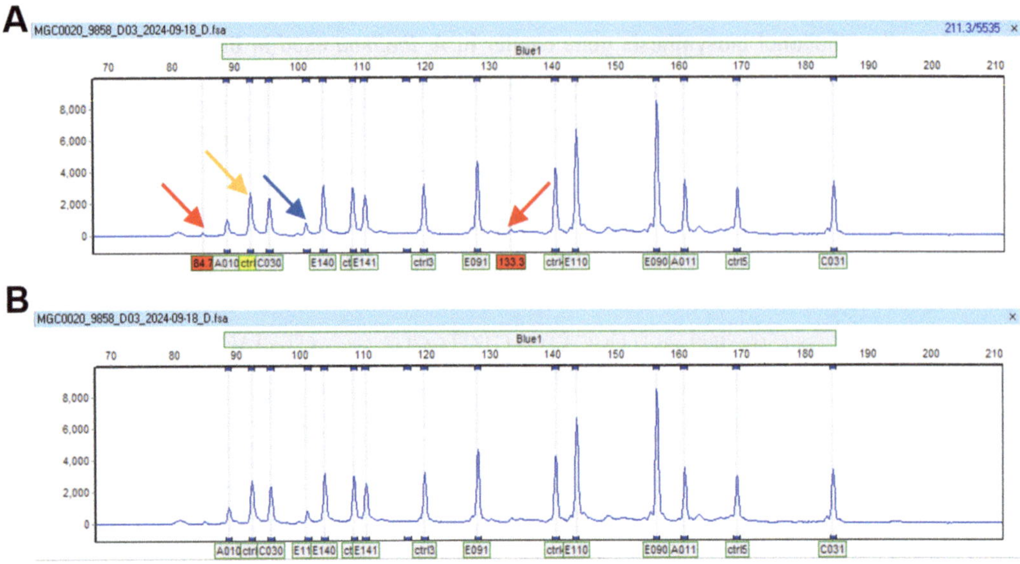

Fig. 4 Manual correction of called peaks. (**a**, **b**) Before (**a**) and after (**b**) the correction. Arrows indicate falsely called alleles—not in the Probe Set Mix, which should be deleted (red), true alleles which require manual confirmation (yellow), and peaks not identified by the automatic analysis, which should be manually added (blue)

7. Select each sample one by one in the *Allele Call* explorer box, visually inspect the graph, and manually correct the allele calls when needed (Fig. 4):
   (a) Check the baseline, peak intensity, and distribution pattern in the control and sample fluorescence channels. Exclude samples with apparent abnormalities by selecting the sample with the right mouse click and choosing "*Disable*" option from the menu. The data for this sample will disappear from the Report.
   (b) If necessary, add missing peaks by right mouse clicking on the peak and selecting *Insert Allele.*
   (c) Inspect all peaks which have been automatically labeled by the software as "*Check*" and "*Undetermined*". Confirm or delete the respective alleles, as necessary, with the right mouse click. The presence of additional alleles, not listed in the Probe Panel, does not affect the subsequent analysis, however we recommend deleting them, for the clarity of the MLPA analysis report.
   (d) Some versions of GeneMarker software may label the saturated peaks as absent ("0"). If this is the case, manually change the status of such peaks to present (*see* **Note 11**).

8. Use *MLPA Analysis* option in the GeneMarker software. Select *Applications* → *MLPA analysis* from the menu, choose "MLPA Ratio" method and check "*Adjust by Control Probes*" check box. You may optionally customize the ratio thresholds for detecting duplications and deletions (*see* **Note 12**). The software will calculate ratios using one automatically selected sample as a Reference Sample. In the next step you may change this selection to any other sample, which should be considered a Reference Sample. A Reference Sample should be well validated and should have all investigated regions (Probe peaks) present in basic copy number. For plants, an accession with a reference genome, e.g., Col-0 in case of *A. thaliana*, is usually a good Reference Sample. An alternative option to select a synthetic reference sample is also available. A synthetic reference is calculated as the median of all selected samples. The program will calculate and display peak intensity/area ratios in the samples compared to the chosen reference.
9. At this stage you may export the results to a file using *Report window* displayed next to the sample ratio plot window. However, in some cases, e.g., when there is no appropriate Reference Sample, automatic MLPA analysis may provide suboptimal results (Fig. 5). In such cases, customized downstream analysis should be performed (*see* **Note 13**).

3.5.2 Data Analysis with Coffalyser.Net

Coffalyser.Net is a free MLPA analysis software developed by MRC Holland, which may be adopted for the analysis of data generated with a custom Probe Set Mix. It assigns peaks to probes based on the fragment size and performs data normalization, quality checks, and comparative sample analysis. The program may be downloaded from the developer's website (https://www.mrcholland.com/) and requires a free license. You need to install the software and set the client connection to a database hosted on the SQL server. Follow the software Installation Manual to set up your working environment. We briefly present how to perform the main steps of the data analysis for MLPA with synthetic Probe Set Mix using Coffalyser.Net client version v.240129.1959. Read the respective Reference Manual for the full description of the program version you use and its options.

1. Run Coffalyser.Net software. Many options in this program are called by the right mouse click on the active window or its element.
2. Go to the folder *CE Devices* in the explorer and click to select the appropriate instrument (e.g., ABI-3130XL) and filter (e.g., G5) used during sequencing.
3. Create a worksheet for your assay or go to the next step if the worksheet has been created already. To create a worksheet, click

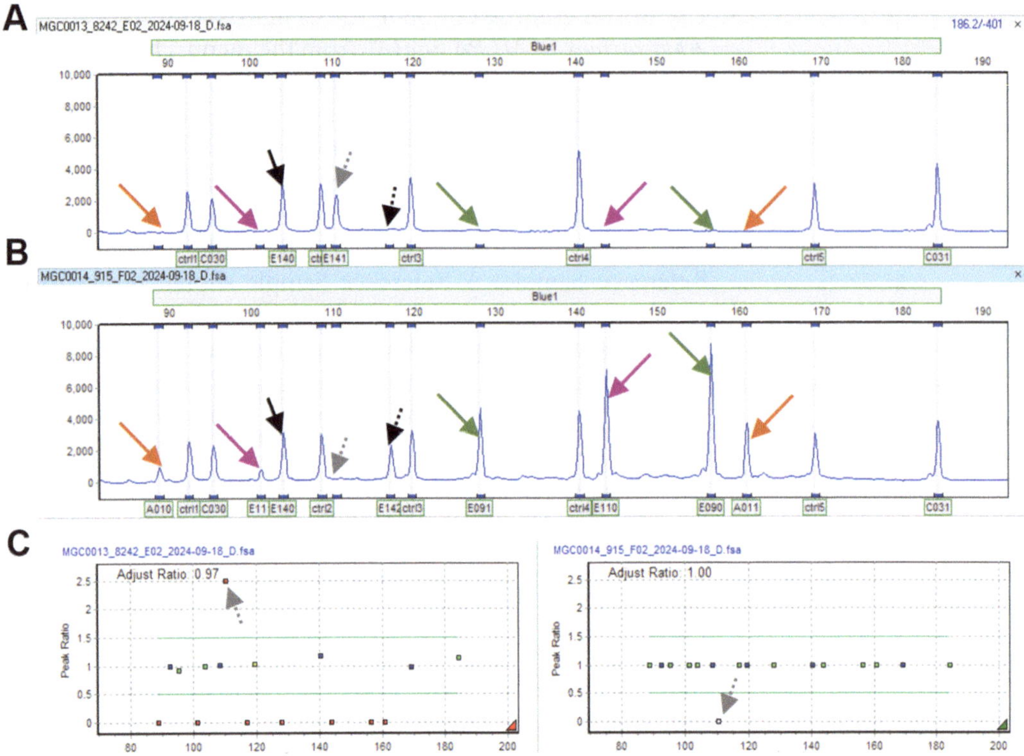

Fig. 5 Comparative sample analysis in GeneMarker. (**a**, **b**) Electropherograms of two *A. thaliana* samples, MGC0013 and MGC0014, after intra-sample normalization. Colored solid arrows indicate peaks for genes present in sample MGC0014 but absent from MGC0013: *AT5G47990* (green), *AT5G48000* (red), and *AT5G48010* (purple). Black solid arrow indicates peak for gene *AT5G47950*, which is present in both samples. Probes marked by dotted arrows, E141 (grey) and E142 (black), detect alternative sequence variants of gene *AT5G47950*, associated with the presence or lack of its inversion [10, 11]. Consequently, only one of the respective peaks may be detected in each sample. (**c**) Ratio plots generated by MLPA analysis module of GeneMarker software, using MGC0014 as a Reference Sample. All inter-sample normalized ratios equal 1 in the Reference Sample, except for E141, which has not been detected, therefore its reference value has been set to 0. In the comparative analysis, E141 peak present in sample MGC0013 has been erroneously interpreted as copy gain. Performing customized analysis is recommended in such cases

the folder *Sheet Library* and select *Open*. The *Coffalyser Work Sheet* Manager will open. Click and select *Add*, then *Create an empty worksheet*. This will generate an empty worksheet, where you may insert all information about the Probes in the assay (*see* **Note 14**).

4. Create a new project: go to the folder *Projects*, click to select *Add Project*, insert *title*, and click *OK*.
5. Click on your project and select *Add experiment*. Insert title and press *OK*. The Experiment window will open.
6. In the *details* tab, select experiment type (*DNA/MLPA*), CE device, dyes, and the worksheet for your sample channel. Then click *Next*.

7. In the *fragment analysis* tab click *Add (From File)* or *Add (Folder)* to import your sample data. If your sample set includes positive reference samples or negative controls, mark them in the *sample type* column accordingly. Then press *Start Fragment Analysis.* If you run a brand new assay, use *auto bin (design length)* probe recognition method at this stage to detect Probe peaks.

8. After fragment analysis, evaluate the data and correct the bins manually, if necessary. To do this, click on any sample and select *Open*. The *Sample Results Explorer* will appear, where you may switch between the samples and different view tabs. Evaluate the *genomic profile* tab. Check for any unexpected peak assignments or the presence of unnamed peaks in the samples (Fig. 6). If you observe such peaks, go to the *binning profile* tab and press *Open Manual Bin Set*. Allow the software to replace the manual bin set setting with the auto bin values when prompted. Then inspect the upper and lower bounds in the suspicious regions and manually correct them if necessary. Make sure the bins for different Probes do not overlap and are not wider than 4 nt. Save your manual corrections to the worksheet (*see* **Note 15**).

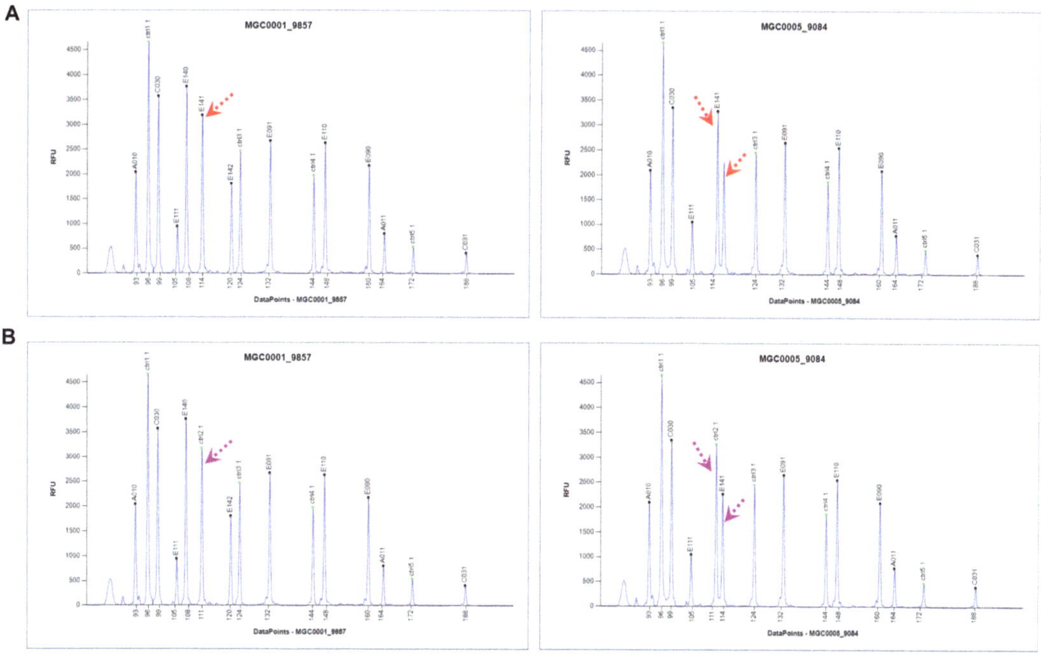

Fig. 6 Manual bin correction in Coffalyser.Net. (**a**) Samples after auto bin analysis; (**b**) the same samples re-analyzed after manual bin correction. Arrows indicate problematic peaks (red) and their corrected assignments to the expected Probe lengths (purple)

9. If manual bin values were created, re-analyze the samples with the new bin settings: press *Start Fragment Analysis* but use *manual* probe recognition method this time.
10. Following fragment identification, go to the *comparative analysis* tab. Select all samples suitable for analysis (*see* **Note 16**) and run comparison. Here you may change the settings regarding peak quantification (e.g., peak height or area) and variation detection ratios (*see* **Note 12**). Press *Start Comparative Analysis*. The program will analyze the data and report warnings if problems were identified for any samples. The results of the analysis are normalized final ratios for successfully processed samples.
11. Click on any sample and select *Open Experiment Results*. From there you may view the data, analysis statistics, and chart statistics for all samples (Fig. 7). If you want to view or save the data for a single sample, right mouse click on it and select *Open Sample Results* option.
12. In the *ratio overview* tab select the type of quantification data you want to be displayed: right mouse click, then select *Show Probe Ratio Type* and choose the desired option. The displayed data may be copied by pressing *Copy Grid Data to Clipboard* button in the main menu, pasted to text editor, and saved to a file. You may copy results from different analysis steps this way, which will also allow you to perform customized normalization and analysis of your data in the external calculation software, if needed (*see* **Note 17**).

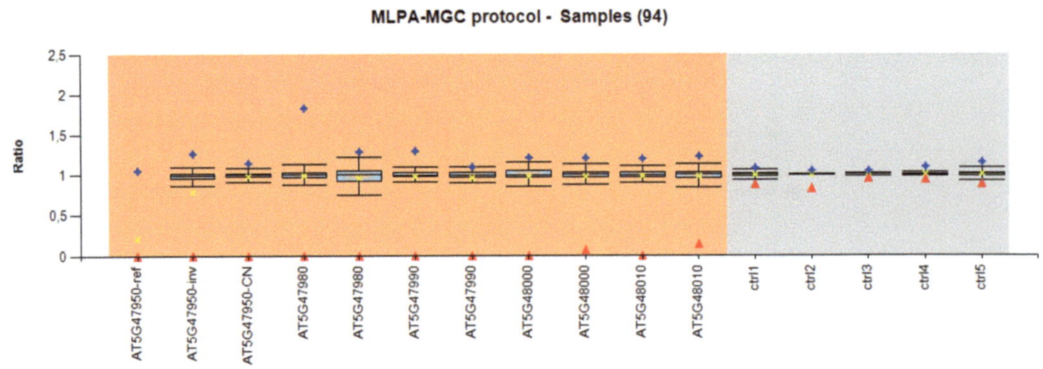

Fig. 7 Statistics chart for *A. thaliana* samples analyzed in Coffalyser.Net. Marks indicate minimum (red triangle), median (yellow cross), and maximum (blue diamond) ratio values across samples. Chart indicates no changes of any control genes in the analyzed set (94 samples). For each gene within thalianol biosynthesis gene cluster, most samples presented no changes, but some gene deletions were also detected. Duplication was also detected with one Probe for gene *AT5G47980*, in a single sample. *AT5G47950* gene inversion, detected by Probe E142, was detected in the majority of samples, while the reference-like orientation was less frequent, which is indicated by the distribution boxplots. Note that no reference sample has been provided, therefore inter-sample normalization based on all samples has been applied. Accordingly, the presented ratios may not directly indicate the gene copy number values

4 Notes

1. Individual PCR tubes may be used instead of PCR tube strips, depending on the number of samples in the batch. However, when processing ≥24 samples, we always use 8-strip PCR tubes, as it reduces sample manipulation time during each step of the MLPA assay. We do not recommend using 96-well PCR plates.
2. The more individual genomes may be searched and compared, the more confident your TSS predictions will be. For some species, it is not necessary to perform multiple sequence alignments. Instead, sequence variation data may be directly extracted from the genome browser of the respective database or from.vcf files generated as a part of the large-scale sequencing projects. For example, for rice you may use SNP-Seek portal which stores data from the analysis of ~3000 rice genomes (https://snp-seek.irri.org/) to directly search for SNPs in your gene of interest [21]. For maize, the information about the sequence variation may be accessed via the Maize Genome Database (https://maizegdb.org/), using JBrowse, GBrowse, or SNPVersity 2.0 tools, all of which display SNPs detected in various maize accessions [22]. For *A. thaliana*, information about small sequence polymorphisms may be extracted from. vcf files, either from the 1001 Genomes Sequencing Project (variants based on short read genomic sequencing of 1135 accessions) or from the Arabidopsis Pangenome Project (variants based on the de novo assemblies generated by long read sequencing of 69 accessions) [23, 24]. Additionally, information about the gene copy number changes among the 1060 *A. thaliana* accessions are available in the AthCNV database (athcnv.ibch.poznan.pl) [18].
3. If it is not possible to place both TSSs in the region completely free of SNPs, minimize the number of overlapping SNPs and make sure that the SNP position is at least 6 nt away from the ligation site. In other terms, avoid any SNPs in the 3′ end of the left TSS and in the 5′end of the right TSS. The presence of SNP at the ligation site or around it will affect the ligation step and may lead to lowering the Probe signal or even result in its complete absence—this in turn may be falsely interpreted as the DNA deletion.
4. SNPs and short mismatches may not be efficiently detected by MLPA. Therefore, the Probes should be tested for their sensitivity and specificity using template DNA with the known genotype at the investigated region. It is also beneficial to include two Probes targeting the same genomic position in your assay: one detecting the reference allele and another

detecting the alternative allele. Each Probe will competitively bind to one sequence type only, increasing the assay specificity. As an internal control, another Probe targeting the same gene but located in another region, free of polymorphisms, may be additionally designed.

5. It is a good practice to create a Probe list for each assay based on Table 1, and add each Probe successively to it. This will prevent duplicating Probe length when designing. We start from adding all the necessary Control Probes to the assay and then fit the newly designed Probes in the remaining slots in the Probe list. For *A. thaliana* we always use the same five Control Probes targeting copy number-invariable genes located on chromosomes 1, 2, 4, and 5 [8]. These Control Probes are 96, 111, 124, 144, and 172 nt in length, which covers the Probe length range of our typical assays. We do not produce any other Probes in these sizes, therefore we may include our Controls Probes in any MLPA experiment for *A. thaliana*.
6. If >250 samples will be analyzed, we recommend to scale up this step and prepare enough Probe Set Mix for all samples in the project. This will minimize technical variation. Divide the Probe Set Mix into aliquots and store them at −20 ° C for later use. Adjust the volume of each aliquot, to minimize the number of freeze-thaw cycles to 6 or less. We successfully use the Probe Set Mix after several years of storage in these conditions. In case we need to prepare an additional amount of the Probe Set Mix or whenever we start a new batch of the MLPA reagents, we re-assay a few previously analyzed samples, to ensure that we obtain identical results.
7. The assay is robust to DNA input, although very low DNA amount may lead to increased variation. For *A. thaliana*, with a diploid genome of ~140 Mb in size, we successfully used from 2 to 100 ng per sample per assay with similar results [17]. For human samples (diploid genome of 3.2 Gb size), 100–250 ng DNA per sample per assay has been used [19, 20]. For species with extremely large genomes (e.g., some plants), it may be beneficial to increase the DNA amount. If you are unsure what DNA amount to use, perform optimization tests for a small set of samples using different DNA inputs, each in at least 3 replicates. Choose the lowest DNA amount resulting in good peak intensity and low sample-to-sample variation.
8. Whenever you open the 8-tube PCR strips, always replace the caps with the new ones, in order to avoid any cross-contamination. We also recommend opening only one strip at a time to reduce the risk of pipetting errors.
9. If you analyze the samples for a new assay and the respective Panel has not been created yet, select "NONE" in the *Panel*

menu. The program will attempt to identify all peaks automatically. The respective alleles will be given numerical names, according to the peak size. You may then use these initial predictions to create a new MLPA Panel in the Panel Editor. Select *Tools* → *Panel Editor* from the menu to open the Editor. There you can select and de-select the samples to visualize electropherogram data, create a table with the description of each Probe, based on the preliminary identified peaks, and save it under a new name. To achieve this, select *File* → *Create New Panel* and choose *Automatically Create* option. Then you may add/remove/rename automatically identified alleles (peaks), based on the expected lengths of Probes in your Probe Set Mix. Alternatively, if another Panel with Probes of similar size already exists, you may select this Panel in the Panel Editor, modify its content to fit the current assay, and save it under a new name. In the next step, you need to re-run the analysis of all samples, only this time select the newly created Panel in the *Run Wizard* dialog box. You may notice small differences between the expected fragment length and allele size assigned by the software based on the internal lane standard (in our assays there is always 3–4 nt discrepancy). That's not a problem as long as these differences consistently affect all peaks and the Probes may be unambiguously identified, based on their relative positions.

10. When creating or editing a Panel, mark all your Control Probes as internal controls, by setting "1" in the Control column in the *Panel Editor*.

11. Peak saturation may occasionally occur in samples with untypically high gene copy number. However, if the peaks are saturated in most samples, try diluting the sample before the capillary electrophoresis. Keep in mind that if the intensity peak has been saturated, the DNA copy number of the respective region may be underestimated.

12. The default ratio options for copy number change detection (<0.75 and >1.30 in GeneMarker, <0.70 and 1.30 in Coffalyser.Net) are recommended for non-selfing species. For selfing species, such as *A. thaliana*, where the heterozygous state is unlikely, more stringent thresholds may be chosen (e.g., <0.5 and >1.5).

13. Sometimes there are no Reference Samples in your set and the Probe Set Mix includes Probes that cannot be inter-sample normalized. For example, there are two Probes detecting alternative alleles of the same region in the assay and only one of them produces peaks in each sample. In such cases we recommend to export the bin data directly after intra-sample normalization, without running automatic analysis. Go to *Report*

Panel on the right, which displays a table of the normalized peak signal for all Probes and all samples. You may adjust the *Report* settings, e.g., by choosing to display peak intensity and/or peak area as the quantification estimate. Copy or save the table into Excel file. Next, perform customized analysis, according to your assay design needs.

14. Coffalyser.Net is optimized for assays designed by MRC Holland and for the human genome. When creating a new worksheet for a different species and synthetic probes, it is sometimes necessary to fill the fields specific for MRC Holland products. We usually fill in the fields in the *details* tab with the assay name or "NA" comment. In the *probes* tab, we fill at least four columns:
    - *gene* column—we provide target gene names and/ or Probe names.
    - *function(s)* column—we select *probe* for all investigated genes and additionally *reference probe (C.N.)* for our Control Probes.
    - *length (design)* column—we provide the expected Probe lengths.
    - *order* column—this is optional but very useful. We order the probes according to their chromosomal location to get the linear view of the entire MGC profile in the samples in the charts and reports.

    The remaining columns may be left empty.
15. In most cases, Coffalyser.Net accurately detects and identifies Probes. Sometimes we need to manually correct fragment bins, e.g., in case when two Probes with similar lengths alternatively bind to the same region and only one of them generates peak in each sample (Fig. 6). If you encounter some problems with saving your manual bin settings from the Manual Bin Set level, insert manual bin set values for all Probes in the respective column of the worksheet and save it. After saving the worksheet with the manual bins added, it is necessary to repeat fragment analysis of all samples, only this time using *manual* probe recognition method. This should fix most peak identification problems.
16. Coffalyser.Net performs a set of automatic quality checks and reports the results in FRSS and FMRS columns. Note that these checks may not work properly when a custom Probe Set Mix is used, e.g., due to the lack of the control fragments expected by the software. We select the samples for subsequent analysis based on manual inspection of the peak profiles.
17. Different types of data may be saved, depending on the user preference. You may, e.g., view and copy *Final Ratios*. These

are the final ratios from the comparative analysis against the reference samples (or the median of all samples if no reference sample has been provided). Alternatively, if you need to perform customized comparative analysis, you may select and copy *Intra Normalized Ratio*, i.e., data that have been normalized within each sample using the Control Probes, but without inter-sample normalization. You may also copy *Probe Height* or *Probe Area* quantification data and perform the normalization and analysis in another program using custom formulas.

Acknowledgments

V.P. has been supported by a scholarship from Charitable Fund in honor of Bizhan Sharopov, the Hero of Ukraine (Ukraine). A.Z. received support from National Science Center (Poland), grant No. 2024/53/B/NZ2/02784. Figure 1: Created in BioRender. Zmienko, A. (2024) BioRender.com/e72x790. Figure 2: Created in BioRender. Zmienko, A. (2024) BioRender.com/z53 k395. Figure 3: Created in BioRender. Zmienko, A. (2024) BioRender.com/s941049.

References

1. Brakhage AA (2013) Regulation of fungal secondary metabolism. Nat Rev Microbiol 11(1): 21–32
2. Nützmann HW, Scazzocchio C, Osbourn A (2018) Metabolic gene clusters in eukaryotes. Annu Rev Genet 52:159–183
3. Go YS, Lee SB, Kim HJ, Kim J et al (2012) Identification of marneral synthase, which is critical for growth and development in Arabidopsis. Plant J 72:791–804
4. Bai Y, Fernández-Calvo P, Ritter A et al (2021) Modulation of Arabidopsis root growth by specialized triterpenes. New Phytol 230:228–243
5. Qi X, Bakht S, Leggett M, Maxwell C et al (2004) A gene cluster for secondary metabolism in oat: implications for the evolution of metabolic diversity in plants. Proc Natl Acad Sci USA 101:8233–8238
6. Shang Y, Ma Y, Zhou Y et al (2014) Biosynthesis, regulation, and domestication of bitterness in cucumber. Science 346:1084–1088
7. Frey M, Chomet P, Glawischnig E et al (1997) Analysis of a chemical plant defense mechanism in grasses. Science 277:696–699
8. Field B, Osbourn AE (2008) Metabolic diversification–independent assembly of operon-like gene clusters in different plants. Science 320:543–547
9. Zhan C, Shen S, Yang C et al (2022) Plant metabolic gene clusters in the multi-omics era. Trends Plant Sci 27:981–1001
10. Liu Z, Cheema J, Vigouroux M et al (2020) Formation and diversification of a paradigm biosynthetic gene cluster in plants. Nat Commun 11(1):5354
11. Marszalek-Zenczak M, Satyr A, Wojciechowski P et al (2023) Analysis of Arabidopsis non-reference accessions reveals high diversity of metabolic gene clusters and discovers new candidate cluster members. Front Plant Sci 14: 1104303
12. Ji W, Osbourn A, Liu Z (2024) Understanding metabolic diversification in plants: branchpoints in the evolution of specialized metabolism. Philos Trans R Soc Lond Ser B Biol Sci 379:20230359
13. Huang AC, Jiang T, Liu YX et al (2019) A specialized metabolic network selectively modulates Arabidopsis root microbiota. Science 364(6440):eaau6389
14. Shirai K, Matsuda F, Nakabayashi R et al (2017) A highly specific genome-wide association study integrated with transcriptome data

reveals the contribution of copy number variations to specialized metabolites in Arabidopsis thaliana accessions. Mol Biol Evol 34(12): 3111–3122

15. Schouten JP, McElgunn CJ, Waaijer R et al (2002) Relative quantification of 40 nucleic acid sequences by multiplex ligation-dependent probe amplification. Nucleic Acids Res 30(12):e57
16. Zmienko A, Samelak-Czajka A, Kozlowski P et al (2016) Arabidopsis thaliana population analysis reveals high plasticity of the genomic region spanning MSH2, AT3G18530 and AT3G18535 genes and provides evidence for NAHR-driven recurrent CNV events occurring in this location. BMC Genomics 17(1):893
17. Samelak-Czajka A, Marszalek-Zenczak M, Marcinkowska-Swojak M et al (2017) MLPA-based analysis of copy number variation in plant populations. Front Plant Sci 8:222
18. Zmienko A, Marszalek-Zenczak M, Wojciechowski P et al (2020) AthCNV: a map of DNA copy number variations in the Arabidopsis genome. Plant Cell 32:1797–1819
19. Marcinkowska-Swojak M, Klonowska K, Figlerowicz M et al (2014) An MLPA-based approach for high-resolution genotyping of disease-related multi-allelic CNVs. Gene 546: 257–262
20. Budzko L, Marcinkowska-Swojak M, Jackowiak P (2016) Copy number variation of genes involved in the hepatitis C virus-human interactome. Sci Rep 6:31340
21. Fuentes RR, Chebotarov D, Duitama J et al (2019) Structural variants in 3000 rice genomes. Genome Res 29:870–880
22. Woodhouse MR, Cannon EK, Portwood JL 2nd et al (2021) A pan-genomic approach to genome databases using maize as a model system. BMC Plant Biol 21:385
23. 1001 Genomes Consortium (2016) 1,135 genomes reveal the global pattern of polymorphism in Arabidopsis thaliana. Cell 166:481–491
24. Lian Q, Huettel B, Walkemeier B et al (2024) A pan-genome of 69 Arabidopsis thaliana accessions reveals a conserved genome structure throughout the global species range. Nat Genet 56:982–991

Chapter 6

Synteny Analysis and Visualization of Gene Clusters

Chanchan Liu, Dishuai Li, and Qinan Wu

Abstract

As the cost of genome sequencing continues to decline, an increasing number of plant genomes are being sequenced and assembled. This progress allows for the comparison of synteny and microsynteny among genomes and gene clusters across various species, facilitating the exploration of their evolutionary relationships.

In this chapter, we describe current methods for analyzing collinearity and synteny among multiple genomes using existing tools from JCVI. We also present a detailed case study that illustrates these methods, utilizing data from four plant genome datasets.

Key words Genome comparison, Synteny analysis, Gene cluster

1 Introduction

1.1 The Role of Synteny and Collinearity Analysis in Species Evolution Studies

Synteny analysis involves comparing the order and sequence similarities of genes across different species or within a single species to identify conserved genomic regions known as syntenic blocks [1]. These conserved regions provide crucial insights into genome structure, evolutionary history, and functional genomics. The core principle of MCscan is based on identifying syntenic regions, either through orthology across different species or through duplications within a species, by analyzing the order and sequence similarities among genes. This process examines patterns of annotated genes within genome assemblies to detect regions with significant sequence conservation [2].

MCscan is particularly advantageous in plant genomics, where frequent genome duplications can obscure evolutionary relationships. Some genome duplication events may precede or follow species divergence, resulting in multiple homologous regions that form syntenic blocks. The identification of these blocks enables researchers to date duplication events, trace the origins of genes (e.g., genome duplicates, tandem duplicates, or single transpositions), and investigate karyotype conservation [3, 4]. Beyond

Hans-Wilhelm Nützmann and Ramesha Thimmappa (eds.), *Metabolic Gene Clusters: Methods and Protocols*,
Methods in Molecular Biology, vol. 3054, https://doi.org/10.1007/978-1-0716-5384-5_6,

enhancing our understanding of evolutionary processes, synteny analysis has practical applications across several domains. In evolutionary biology, it allows researchers to trace gene origins and elucidate genome evolution across species. In gene function prediction, the identification of syntenic blocks aids in pinpointing conserved gene clusters with similar functions, providing crucial insights into the roles of previously uncharacterized genes [5]. Additionally, in crop improvement, synteny analysis can guide the identification of genes associated with important agronomic traits, thereby facilitating the development of superior crop varieties [6].

By leveraging the principles of synteny analysis, MCscan not only offers a detailed map of genomic conservation and divergence but also serves as a powerful tool for advancing our understanding of gene function and enhancing agricultural practices.

1.2 The Discovery and Progress of Plant Gene Clusters

Plant gene clusters refer to groups of functionally related genes that are clustered together within the plant genome. These genes typically participate in the same biosynthetic pathway, particularly in the secondary metabolism of plants. Secondary metabolites include various bioactive substances, such as antimicrobial compounds, pigments, and volatile oils, which are crucial for the survival and reproduction of plants. These compounds can be used to fend off pathogens, attract pollinators, or serve as defense mechanisms against herbivores. While this pattern of gene organization has been extensively studied in microorganisms, its investigation in plants has only emerged as a significant field of study in recent years [7, 8].

Identification of Gene Clusters: With the advancement of genome sequencing technologies and bioinformatics, scientists can now identify and annotate gene clusters within plant genomes through high-throughput sequencing, bioinformatic analysis, genome-wide association studies (GWAS), functional genomics, and metabolomics. The application of these techniques allows researchers to explore the connections between gene clusters and specific biochemical pathways at the genome-wide level, thereby revealing their biosynthetic networks [9].

Evolution and Ecological Function of Gene Clusters: The formation of gene clusters is closely related to recombination and duplication events within plant genomes. During evolution, factors such as gene duplication, transposition, and selective pressures drive the formation and rearrangement of gene clusters, enabling plants to adapt to specific environmental conditions. From an ecological perspective, gene clusters enhance plant adaptability to environmental changes by synthesizing specific chemical compounds, such as secondary metabolites, to respond to biotic and abiotic stresses [10, 11].

Biotechnological Applications of Gene Clusters: Understanding the structure and function of gene clusters is of significant value in biotechnology. Scientists can manipulate gene clusters through genetic engineering or synthetic biology to optimize and enhance the biosynthetic capabilities of plants, thereby producing compounds with commercial or medicinal value. For example, by overexpressing transcription factors or employing gene editing technologies, it is possible to regulate the activity of specific gene clusters, thereby increasing the yield of certain secondary metabolites [12].

1.3 Comprehensive Guide to Synteny Analysis of Gene Clusters

In this chapter, we present a detailed method for synteny analysis of gene clusters. Before starting the analysis, it is crucial to have fully assembled and annotated genomic data, including a CDS sequence file and a GFF file derived from chromosome assembly. The quality of the genome assembly directly affects the reliability of synteny analysis, as incomplete assemblies can lead to disorganized synteny blocks, making visualization and cross-species comparisons difficult. Therefore, it is recommended to use the most complete genome data available.

To access the required data, databases such as NGDC and NCBI are useful resources. After obtaining the datasets, renaming files and adjusting formats may be necessary to ensure compatibility with analysis software. During synteny analysis, orthologous sequences between species are compared, and visualization tools are used to highlight the synteny blocks on chromosomes. Finally, gene-level analysis is performed, allowing for a detailed examination of gene cluster collinearity across species.

This approach facilitates a comprehensive understanding of the evolutionary relationships and functional organization of gene clusters, providing valuable insights for further research.

2 Methods

2.1 Preparing the Genomic Dataset

Before initiating collinearity analysis, it is essential to prepare comprehensive genomic data, including a properly assembled and annotated CDS sequence file (cds.fasta) and a GFF file derived from the chromosome assembly. Collinearity analysis demands a high-quality genome assembly. In some genomes, the successful assembly of genes into their respective chromosomal locations is limited, resulting in poor scaffold integrity and the presence of numerous contigs. Therefore, it is crucial to prioritize the use of more completely assembled genomic data before analysis. Otherwise, the synteny block data generated subsequently may be disorganized, rendering visualization difficult and hindering synteny block analysis across multiple species. Similarly, for diploid genome assemblies yielding two genome sets, the set with more complete chromosome assembly should be prioritized for subsequent collinearity analysis.

In the National Genomics Data Center (NGDC) database, we first search for the species' Latin name, e.g., *Schizonepeta tenuifolia*, through the website. Under the "Database" section, select "GWH (Genome Warehouse)", where currently three records can be found. This indicates that there are three relevant whole-genome datasets available for this species. Clicking on the number "3" leads to the search results page. Next, locate the genome information for *Schizonepeta tenuifolia* by checking the species name, and by clicking on the corresponding entry, access the detailed genome information page. On this page, various details of the genome can be viewed. In the "Download" section, there is a button for downloading the DNA sequence. Additionally, based on the files provided by the uploader, download links for genome functional annotation files, transcriptome assembly files, and protein sequence files are also available.

In the NCBI database, first search for the species' Latin name, *Schizonepeta tenuifolia*, through the website. Under the "Genomes" section, locate the "Assembly/Genome" project and click on it to view all the available genome information for this species. NCBI marks the recognized genome as the reference genome. Then click on the Assembly code ASM2949068v1 for *Schizonepeta tenuifolia*, where detailed information about the genome can be found. Upon clicking "Download," options for selecting the download source appear. RefSeq provides high-quality, validated reference sequences; GenBank contains a broader range of sequence data, suitable for researchers needing raw sequence data or more extensive datasets; the "All" option includes data from both RefSeq and GenBank, offering the most comprehensive search results.

For file types, the options include: Genome sequences (FASTA), which contain only the DNA sequences of the genome without annotation information; Annotation features (GTF), which include annotations of genes and other functional elements on the genome, commonly used for gene expression analysis; Annotation features (GFF), which are similar to GTF but with slight format differences, also used for genome annotation; Sequence and annotation (GBFF), which is a comprehensive file containing both genome sequences and complete annotation information; Transcripts (FASTA), which contain sequences of gene transcripts, used for studying gene expression; Genomic coding sequences (FASTA), which contain only the DNA sequences encoding proteins (CDS); Protein (FASTA), which contains the translated protein sequences; Sequence report (JSONL), which is a statistical report of the sequences, suitable for programmatic processing; Assembly data report (JSONL), which contains detailed data on genome assembly, also suitable for programmatic analysis. Of course, the selection of downloadable files largely depends on the data provided by the uploader. After selecting the required files and downloading them, you can find the assembled genome files of *Schizonepeta tenuifolia* in the ncbi_dataset folder after unzipping the package.

After downloading the genome dataset, the files can be renamed to facilitate subsequent analysis. Due to inconsistencies in sequencing and assembly formats from different laboratories, it may be necessary to adjust the data format to meet the requirements of the analysis software. For example, in collinearity analysis, we might encounter a situation where the gene names extracted from the GFF file are inconsistent with those in the CDS file. Therefore, we need to modify the gene names in the CDS file.

For example, change ">lcl|NC_080387.1_cds_XP_057790285.1_1 [gene=LOC131007145] [db_xref=GeneID:131007145] [protein=transcription termination factor MTERF9, chloroplastic-like isoform X1] [protein_id=XP_057790285.1] [location=join(47905..48000,48945..50069)] [gbkey=CDS]" to ">cds_XP_057790285.1_1". We can use Python to perform this modification:

```
# Open input and output files
with open('mil_cds.fna', 'r') as infile, open('mil_cds_modified.fna', 'w') as outfile:
    for line in infile:
        # If the line starts with ">", it is a header line
        if line.startswith('>'):
            # Extract protein_id from the original header row
                start_index = line.find('protein_id=') + len('protein_id=')
            end_index = line.find(']', start_index)
            protein_id = line[start_index:end_index]

            # Modify the title to the new format
            new_header = f">cds-{protein_id}\n"
            outfile.write(new_header)
        else:
            # Write the non-header rows as is
            outfile.write(line)
```

Next, proceed to convert the GFF file into a BED file. During this process, it is important to check whether the gene names in the GFF file are referenced by ID or Name. If they are referenced by ID, you need to change --key=Name to --key=ID. Additionally, the FASTA file should be reformatted. At this point, we will have obtained the genome dataset necessary for collinearity analysis.

```
python -m jcvi.formats.gff bed --type=mRNA --key=ID Ste.gff -o
ten.bed
python -m jcvi.formats.fasta format CRR707903.fasta.gz ten.cds
```

If the genome annotation dataset contains lots of isoforms, we could add one option to the command line:

```
python -m jcvi.formats.gff bed --type=mRNA --key=ID --pri-
mary_only Ste.gff -o ten.bed
```

With this, the genomic data needed for collinearity analysis is prepared. In this demonstration, we will use *Schizonepeta tenuifolia*, *Mentha longifolia*, *Hyssopus officinalis*, and *Agastache rugosa* as examples.

2.2 Synteny Analysis with MCscan

We first compare the orthologous sequences between the two species. During this process, remove tandem duplications and weak hits.

```
python -m jcvi.compara.catalog ortholog ten rug --cscore=.99
--no_strip_names
```

Subsequently, the following files will be obtained:

```
ten.rug.anchors ten.rug.last.filtered ten.rug.pdf ten.rug.
last ten.rug.lifted.anchors
```

The ten.rug.last file is the raw LAST output, ten.rug.last.filtered is the filtered LAST output, ten.rug.anchors represents the seed synteny blocks (high quality), and ten.rug.lifted.anchors recruits additional anchors to form the final synteny blocks. Since the anchors file serves as the seed for subsequent synteny blocks, it's crucial to pay attention to the order of species during this step, as it will affect the final presentation order of the synteny relationships. In the example presented here, the synteny relationships between *Schizonepeta tenuifolia* and *Agastache rugosa*, *Agastache rugosa* and *Mentha longifolia*, and *Agastache rugosa* and *Hyssopus officinalis* are explored. Therefore, three comparisons are conducted and the anchors files for each pairwise species comparison are obtained.

```
python -m jcvi.compara.catalog ortholog rug lon --cscore=.99
--no_strip_names
python -m jcvi.compara.catalog ortholog rug off --cscore=.99
--no_strip_names
```

Next, the synteny visualization between multiple species' genomes based on the anchors and BED files is analyzed. First, a seqids file is constructed, which informs the plotter which set of main chromosomes to include. Next, remove any scaffold fragments that are too short are removed. Additionally, the order of the chromosomes in this file should match the vertical order of species in the final visualization.

```
HiC scaffold 1,HiC scaffold 2,HiC scaffold 3,HiC scaffold 4,
HiC scaffold 5,HiC scaffold 6
Chr1.1,Chr2.1,Chr3.1,Chr4.1,Chr5.1,Chr6.1,Chr7.1,Chr8.1,
Chr9.1
......
```

Next is the layout file, which tells the plotter where to draw each element.

```
# y, xstart, xend, rotation, color, label, va,  bed
 .8,     .2,    .8,       0,      , ten, top, ten.bed
 .6,     .2,    .8,       0,      , rug, top, rug.bed
 .3,     .05,    .45,   -30,      , lon, bottom, lon.bed
 .3,     .5,    .95,      30,      , off, bottom, off.bed
# edges
e, 0, 1, ten.rug.anchors.simple
e, 1, 2, rug.lon.anchors.simple
e, 1, 3, rug.off.anchors.simple
```

The whole canvas ranges from 0 to 1 on both the x-axis and y-axis. First, the three columns (y, xstart, xend) specify the position of the track. Then, the rotation, color, label, vertical alignment (va), and the path to the genome BED file are defined. Track 0 now corresponds to *Schizonepeta tenuifolia*, Track 1 to *Agastache rugosa*, Track 2 to *Mentha longifolia*, and Track 3 to *Hyssopus officinalis.* Then the following command is used to generate a . simple file, which is a more succinct form than the .anchors file. Additionally, the position, color, labels, and other information in the layout file can be adjusted to optimize the synteny visualization for the best presentation.

```
python -m jcvi.compara.synteny screen --minspan=30 --simple
ten.rug.anchors ten.rug.anchors.new
python -m jcvi.compara.synteny screen --minspan=30 --simple
rug.lon.anchors rug.lon.anchors.new
```

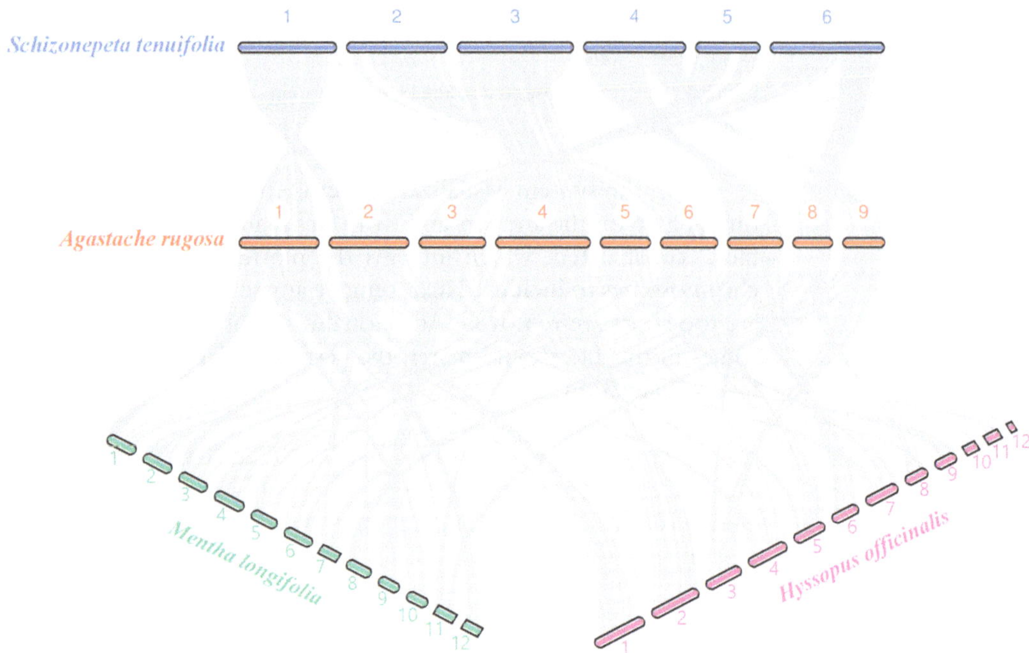

Fig. 1 Macrosynteny analysis showing large-scale genomic differences between *S. tenuifolia*, *A. rugosa*, *M. longifolia*, and *H. officinalis*

```
python -m jcvi.compara.synteny screen --minspan=30 --simple
rug.off.anchors rug.off.anchors.new
```

Then all input files can be plotted using the following command:

```
python -m jcvi.graphics.karyotype seqids layout
```

Based on this analysis, the positional information of synteny blocks on the chromosomes across the four species is obtained (Fig. 1). It is evident which chromosomes the synteny blocks are primarily concentrated on. Additionally, synteny blocks of interest can be highlighted by coloring them in the .simple file for further analysis.

Next, we will focus on the collinearity analysis at the gene level. The following command is used to extract the best matching region for each species from the .anchors file obtained in the previous analysis. The option "--iter=1" means one best region is extracted; if "--iter=2" is set, it means two regions will be extracted.

```
python -m jcvi.compara.synteny mcscan ten.bed ten.rug.lifted.
anchors --iter=1 -o ten.rug.i1.blocks
```

The output file ten.rug.i1.blocks contains numerous aligned gene pairs. In the example, three files will be obtained: ten.rug.i1.blocks, rug.lon.i1.blocks, and rug.off.i1.blocks. Therefore, the next step is to merge these blocks files to visualize the gene cluster collinearity relationships across multiple species.

```
python -m jcvi.formats.base join ten.rug.i1.blocks rug.lon.i1.
blocks rug.off.i1.blocks--noheader | cut -f1,2,4,6 > rug.
blocks
```

When merging the blocks files for collinearity visualization, the following issue often arises: the block file contains regions of collinearity across the entire genomes of multiple species, but these regions may be distributed across different chromosomes, leading to a somewhat disorganized visualization. To address this, the *Agastache rugosa* blocks file is used as a template. After selecting the gene information on the *Agastache rugosa* gene cluster, the VLOOKUP function is used to search the other two blocks files for segments that show collinearity with the *Agastache rugosa* gene cluster region. Aim to ensure that the selected regions are concentrated on a single chromosome in each species. This results in the blocks2 file that contains the regions of interest. Next, a visual analysis of this blocks file is performed, and a new layout file, blocks.layout, contains the following command:

```
# x,   y, rotation,    ha,     va, color, ratio,          label
0.5, 0.5,         0, left,    center,     ,     2,      rug Chr8
0.5, 0.7,         0, left,    center,     ,     2,      ten sca6
0.3, 0.3,        0, center, bottom,     ,     1,      lon sca11
0.7, 0.3,        0, center, bottom,     ,    .5,       off sca8
# edges
e, 0, 1
e, 0, 2
e, 0, 3
```

Then the command is run as follows:

```
python -m jcvi.formats.bed merge ten.bed rug.bed lon.bed off.
bed-o ten_rug_lon_off.bed
python -m jcvi.graphics.synteny blocks2 ten_rug_lon_off.bed
blocks2.layout
```

The plots thus created are shown in Fig. 2.

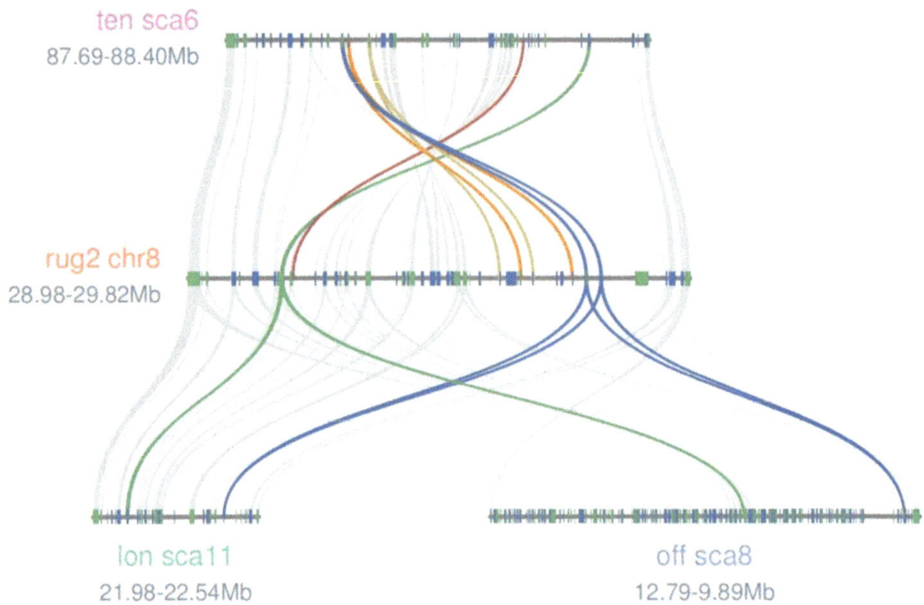

Fig. 2 Collinearity (synteny) analysis of a bipartite BGC compared with other high-quality Lamiaceae genomes

Acknowledgment

This work was supported by NSFC grant 82373978.

References

1. Tang H, Bowers JE, Wang X, Ming R, Alam M, Paterson AH (2008) Synteny and collinearity in plant genomes. Science 320(5875): 486–488. https://doi.org/10.1126/science.1153917
2. Wang Y, Tang H, Debarry JD, Tan X, Li J, Wang X, Lee TH, Jin H, Marler B, Guo H, Kissinger JC, Paterson AH (2012) MCScanX: a toolkit for detection and evolutionary analysis of gene synteny and collinearity. Nucleic Acids Res 40(7):e49. https://doi.org/10.1093/nar/gkr1293
3. Vandepoele K, Simillion C, Van de Peer Y (2003) Evidence that rice and other cereals are ancient aneuploids. Plant Cell 15(9): 2192–2202. https://doi.org/10.1105/tpc.014019
4. Cannon SB, Mitra A, Baumgarten A, Young ND, May G (2004) The roles of segmental and tandem gene duplication in the evolution of large gene families in *Arabidopsis thaliana*. BMC Plant Biol 4:10. https://doi.org/10.1186/1471-2229-4-10
5. Liu C, Smit SJ, Dang J, Zhou P, Godden GT, Jiang Z, Liu W, Liu L, Lin W, Duan J, Wu Q, Lichman BR (2023) A chromosome-level genome assembly reveals that a bipartite gene cluster formed via an inverted duplication controls monoterpenoid biosynthesis in *Schizonepeta tenuifolia*. Mol Plant 16(3):533–548. https://doi.org/10.1016/j.molp.2023.01.004
6. Salse J, Bolot S, Throude M, Jouffe V, Piegu B, Quraishi UM, Calcagno T, Cooke R, Delseny M, Feuillet C (2008) Identification and characterization of shared duplications between rice and wheat provide new insight into grass genome evolution. Plant Cell 20(1):11–24. https://doi.org/10.1105/tpc.107.056309
7. Nützmann HW, Osbourn A (2014) Gene clustering in plant specialized metabolism. Curr Opin Biotechnol 26:91–99. https://doi.org/10.1016/j.copbio.2013.10.009
8. Kliebenstein DJ, Osbourn A (2012) Making new molecules – evolution of pathways for

novel metabolites in plants. Curr Opin Plant Biol 15(4):415–423. https://doi.org/10.1016/j.pbi.2012.05.005

9. Nützmann HW, Huang A, Osbourn A (2016) Plant metabolic clusters – from genetics to genomics. New Phytol 211(3):771–789. https://doi.org/10.1111/nph.13981

10. Hanada K, Zou C, Lehti-Shiu MD, Shinozaki K, Shiu SH (2008) Importance of lineage-specific expansion of plant tandem duplicates in the adaptive response to environmental stimuli. Plant Physiol 148(2):993–1003. https://doi.org/10.1104/pp.108.122457

11. Kliebenstein DJ (2008) A role for gene duplication and natural variation of gene expression in the evolution of metabolism. PLoS One 3(3):e1838. https://doi.org/10.1371/journal.pone.0001838

12. Medema MH, Osbourn A (2016) Computational genomic identification and functional reconstitution of plant natural product biosynthetic pathways. Nat Prod Rep 33(8):951–962. https://doi.org/10.1039/c6np00035e

Chapter 7

Automating Plasmid Construction for High-Throughput Heterologous Expression of Fungal Gene Clusters

Anjali D. Shah, Andrew M. Bailey, and Katherine Williams

Abstract

Fungi are renowned for producing a diverse range of small-molecule natural products with applications in the food, pharmaceutical, and agrochemical industries. While bioinformatics can rapidly identify potential biosynthetic gene clusters (BGCs), their functional analysis remains a bottleneck, limiting the discovery of novel compounds. Advances in molecular techniques and genome sequencing now enable heterologous expression of biosynthetic pathways in hosts like *Aspergillus oryzae*, facilitating access to metabolites not achievable through traditional methods. However, given the high number of uncharacterized gene clusters, a high-throughput process is crucial for efficiently mining novel natural products. This chapter outlines a streamlined, high-throughput protocol using yeast homologous recombination and automated liquid handling to construct multigene plasmids for heterologous expression. While this chapter uses *A. oryzae* NSAR1 and the pTYGS series of expression vectors as an example, the core principles of this robotics-driven protocol are adaptable to other heterologous hosts, vectors, and organisms, providing a powerful tool for high-throughput discovery and engineering of natural products.

Key words High-throughput, Automation, Yeast homologous recombination, Liquid-handling robot, Fungal natural products, *Aspergillus oryzae*, Biosynthetic gene clusters, Secondary metabolites, Heterologous expression

1 Introduction

Fungi are prolific producers of chemically diverse natural products, many of which exhibit a wide range of bioactive properties. Penicillin, produced by *Penicillium* sp. [1] revolutionized medicine by effectively treating bacterial infections. This paved the way for further exploration into microbial sources for other antibiotics and drugs. This effort led to breakthroughs such as the antifungal griseofulvin [2], immune-modulator cyclosporin [3], antibiotic pleuromutilin [4, 5], and the polyketide lovastatin [6, 7], which gave rise to the statin class of cholesterol-lowering medications. However, in light of the ongoing antimicrobial resistance crisis, the search for novel compounds has become more urgent. Fungi

Hans-Wilhelm Nützmann and Ramesha Thimmappa (eds.), *Metabolic Gene Clusters: Methods and Protocols*, Methods in Molecular Biology, vol. 3054, https://doi.org/10.1007/978-1-0716-5384-5_7,

are also important sources of bioactive agrochemicals, for example, the antifungal strobilurins [8, 9]. Despite an estimated 12 million fungal species [10], fungi remain a largely untapped resource. Combined with other sources like bacteria, plants, and algae, and empowered by advances in genome mining and bioinformatics, this resource holds immense potential for discovering new bioactive compounds, including antimicrobials with novel modes of action [11].

While bioinformatics can rapidly identify potential biosynthetic gene clusters (BGCs) encoding the biosynthesis of these potentially bioactive compounds [12–17], their downstream functional analysis is time-consuming and severely constrains the discovery of novel compounds [18]. Often, microbes are challenging to culture and typically produce low titers of secondary metabolites under laboratory conditions. Furthermore, genome mining frequently uncovers numerous gene clusters, creating challenges in prioritization and scalability for effective natural product discovery [19].

This chapter presents a cloning strategy for multigene expression in a heterologous host, paired with automation using a liquid handling robot. By circumventing challenges associated with wild-type fermentations, this approach enables pathway genes to be refactored and expressed in a suitable host strain. The example given below focuses on refactoring fungal biosynthetic gene clusters for heterologous expression in *Aspergillus oryzae*, the fungus used in the production of foods such as soy sauce [20]. However, the general approach outlined here is applicable to a wide range of systems. Expressing various combinations of pathway genes in this way facilitates the dissection of complex biosynthetic pathways, allowing the order of catalytic transformations to be understood. This flexible method also supports pathway engineering through the "mixing and matching" of genes, a cutting-edge technique for generating chemical diversity [21]. Automation further enhances scalability, enabling high-throughput analysis of multiple gene clusters for faster, simpler, and cost-effective identification of novel, natural, and non-natural products.

Apart from *A. oryzae,* other hosts have been utilized for the heterologous production of fungal natural products, such as *Aspergillus nidulans* [22, 23], and *Fusarium* sp. [24]. As mentioned above, the methods discussed in this chapter use *A. oryzae*, a domesticated relative of *Aspergillus flavus*, with a long history of use in food and drink production. It has GRAS ("Generally Recognized As Safe") status and produces relatively few natural products under laboratory conditions, making it an ideal cellular factory for investigation into secondary metabolites from other species [20]. In addition, specific strains of *A. oryzae* with corresponding molecular toolkits have been developed for just that purpose. NSAR1 is a quadruply auxotrophic (*niaD*$^-$ *sC*$^-$ Δ*argB* *adeA*$^-$)

A. oryzae strain, which means that four separate transformation constructs can be selected for simultaneously [25]. The development of four corresponding multigene expression vectors, each of which contains four separate cassettes, means that 16 genes can easily be co-expressed [26].

The methods herein guide the reader through an automated process to generate heterologous expression plasmids containing fungal gene clusters. The method is an advancement on the manual method outlined in our previous chapter [27]. The protocol encompasses plasmid and primer design, gene amplification using polymerase chain reaction (PCR) and construction using yeast homologous recombination, facilitated by the Freedom Evo Tecan liquid handling robot, to create libraries of plasmids with multiple genes of interest (Fig. 1). While tailored for *A. oryzae* NSAR1 and the pTYGS series of plasmids, the overarching principles of this protocol can be adapted for use beyond fungal gene clusters and with other heterologous hosts and vectors.

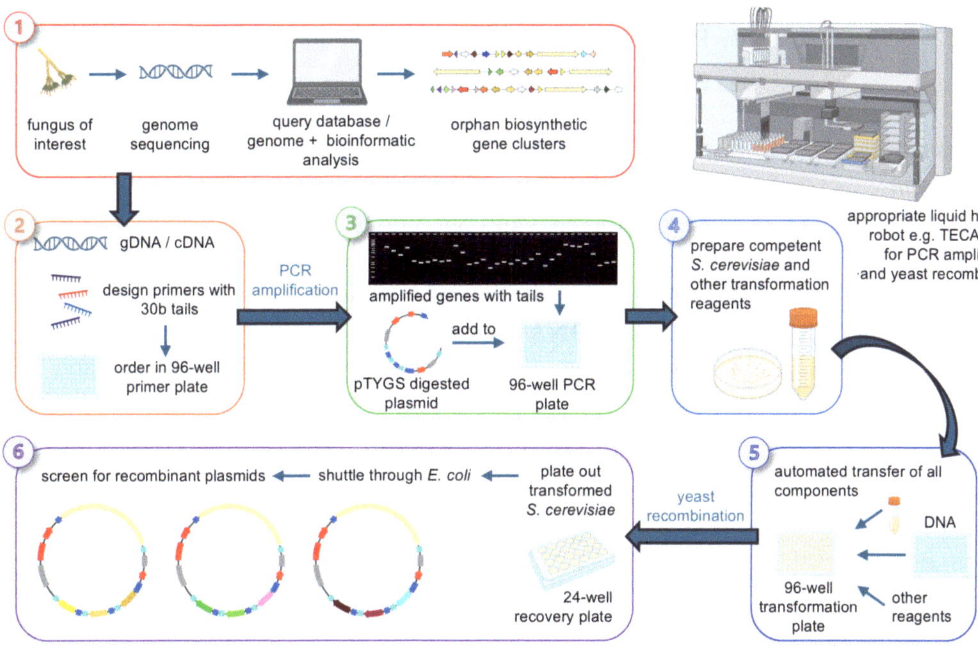

Fig. 1 An overview of an automated process for generating libraries of multigene plasmids for heterologous expression via yeast recombination. *Step 1:* Identification of orphan biosynthetic gene clusters for downstream high-throughput analysis by yeast recombination. *Step 2:* Isolation of genomic DNA or complementary DNA to provide PCR template and primer design including 30 base compound tails for amplification of desired genes. *Step 3:* Automated PCR amplification of all genes and digestion of pTYGS plasmid, prepared in a 96-well PCR plate. *Step 4:* Preparation of competent *S. cerevisiae* and remaining transformation reagents including PEG, LiOAc, and SS-DNA. *Step 5:* Automated transfer of each component to the 96-well transformation plate in the required order and combination. *Step 6:* Automated plating of transformed cells into a 24-well recovery plate and subsequent manual *E. coli* shuttling to screen recombinant plasmids for heterologous expression

2 Materials

2.1 General Laboratory Equipment

1. Liquid handling robot (*see* **Note 1**).
2. Benchtop microcentrifuge for centrifugation of 1.5 mL microcentrifuge tubes.
3. Benchtop centrifuge with a 96- or 384-well plate rotor.
4. Freestanding centrifuge with rotor for 50 mL conical tubes.
5. 1.5 mL microcentrifuge tubes.
6. 50 mL conical centrifuge tubes.
7. Sterile water.
8. Static incubator.
9. Shaking incubator.
10. Heating block or water bath for *Saccharomyces cerevisiae* and *Escherichia coli* transformations.
11. Petri dishes.
12. 250 mL conical flasks.
13. PCR thermocycler.
14. Gel electrophoresis equipment.

2.2 Plasmid and Primer Design

1. pTYGS plasmid series, or other appropriate yeast-*E. coli* shuttle expression vector for use in a suitable host strain.
2. Restriction enzymes *SgsI* and *NotI*.

2.3 High-Fidelity Polymerase Chain Reaction (PCR)

1. High-fidelity polymerase (*see* **Note 2**).
2. Genomic DNA (gDNA) or complementary DNA (cDNA) isolated from native producer or a genetically capable but non-producing strain.
3. Primers in 96- or 384-well plates from synthetic DNA ordering service.
4. 96- or 384-well PCR plates.

2.4 Construction of the Multi-gene Expression Plasmids via Yeast Recombination

1. A uracil-auxotrophic strain of *S. cerevisiae* strain (*see* **Note 3**).
2. YPAD: 1% (w/v) yeast extract, 2% (w/v) peptone, 2% (w/v) glucose, 0.04% (w/v) adenine sulfate, plus 1.5% (w/v) agar for solid medium.
3. PEG solution: 50% (w/v) polyethylene glycol 3350 (*see* **Note 4**).
4. 1 M lithium acetate (LiOAc) and 0.1 M LiOAc.
5. SS-DNA: 2 mg/mL salmon sperm DNA (SS-DNA) in TE (Tris–EDTA buffer solution: 10 mM Tris–HCl, 1 mM Na_2EDTA, pH 8.0). *See* **Note 5** for details on preparation of SS-DNA.

6. Pre-mixed 1 M LiOAc: denatured SS-DNA in 1:1.4 ratio. Store frozen and keep on ice once thawed (*see* **Note 6**).
7. SM-URA agar: 0.17% (w/v) yeast nitrogen base, 0.5% (w/v) ammonium sulfate, 2% (w/v) glucose, 0.077% (w/v) complete supplement mixture minus uracil (Q-biogene), 1.5% (w/v) agar.
8. 96-well plates which can accommodate at least 0.5 mL/well.
9. 24-well plates.
10. Digested pTYGS plasmid series, or other appropriate digested yeast-*E. coli* shuttle expression vector for use in a suitable host strain (from section 3.1).
11. PCR products (from section 3.3).

3 Methods

Identification and annotation of fungal biosynthetic gene clusters of interest is outside the scope of this chapter. To undertake these types of analyses, please see the comprehensive guide in reference [28] for more information. Generation of an annotated biosynthetic gene cluster is required for the protocol outlined below.

3.1 Plasmid Design

As mentioned before, this method utilizes the pTYGS series of multigene expression vectors for expression in *Aspergillus oryzae* NSAR1. While this example is used, the core principles of the automated workflow described below are adaptable to any vector for expression in various heterologous hosts.

The pTYGS series of vectors differ solely in their selectable marker for NSAR1 transformation (either *argB* [29], *adeA* [25], *niaD* [30], or *sC* [30]). Each vector features an *amyB* expression cassette, which is highly inducible in the presence of starch or maltose as the carbon source [31]. Additionally, three further cassettes (*gpdA*, *eno*, and *adh*) ensure strong constitutive expression in *A. oryzae* [32]. The *amyB* cassette includes a Gateway® destination fragment, enabling the expression of large genes that are first assembled in a Gateway entry vector and then inserted by site-specific recombination in vitro [32]. This destination fragment contains the *ccdB* gene, which encodes the toxic ccdB protein, meaning these plasmids require propagation in a ccdB-resistant *E. coli* strain (*see* **Note 7**). Subsequent recombinant plasmid propagation in a susceptible strain effectively selects against the non-recombinant vector.

The pTYGS vectors include the ColE1 origin of replication and the *ampR* gene for replication and selection in *E. coli*. Additionally, they contain the *S. cerevisiae* 2 μm origin of replication [33] and *URA3* gene [34], allowing for the construction of the expression plasmids in yeast. This process exploits the natural ability of *S. cerevisiae* to perform homologous recombination, enabling precise assembly of multiple DNA fragments without the need for unique restriction sites and ligation [35]. Furthermore, recombination can be directed to any homologous region within the DNA fragments, unlike Gibson Assembly, which requires 5′ exonuclease activity to create overlapping sequences for assembly.

1. Choose the entry vector to be used and purchase from a relevant supplier. If using the pTYGS series of vectors, see below for further design details (*see* **Note 8**).
2. Determine if more than one plasmid is required (*see* **Note 9**). If using the pTYGS series of vectors, up to four genes can be expressed from a single plasmid.
3. Design the plasmid, deciding where genes are to be inserted (*see* **Note 10**). If fewer than four genes are required in a single plasmid, there are two possible design options: (a) If empty cassettes are needed for subsequent experiments, a suitable patch can be recombined to repair the cut *SgsI* sites. (b) If empty cassettes are not needed, genes can be inserted in relevant places to remove other cassettes. See Fig. 2 for details.

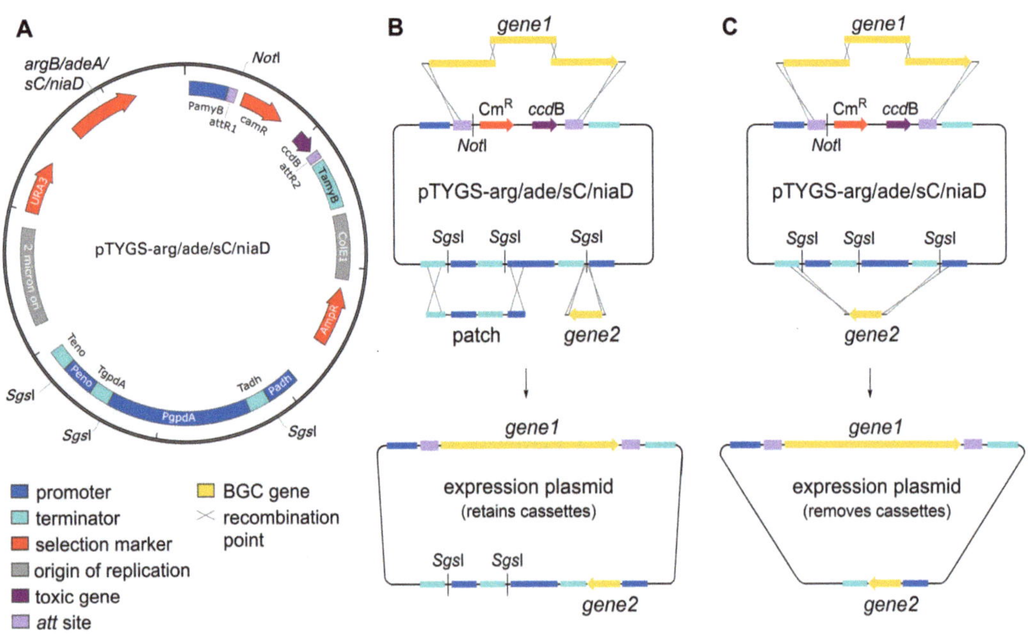

Fig. 2 **(a)** Map of the pTYGS series of vectors. Each version of the vector contains a different complementation marker (*argB/adeA/sC/niaD*) for use with *A. oryzae* NSAR1. **(b)** Plasmid design to retain empty cassettes by insertion of an amplified portion of the plasmid acting as a "patch" to repair two *SgsI* sites. **(c)** Plasmid design to remove empty cassettes by insertion of one gene between the P_{adh} promoter and the T_{eno} terminator

4. If using a plasmid backbone, manually prepare the plasmid of choice by restriction digest according to the manufacturer's instructions (*see* **Note 11**). Check the digestion has completed on a gel and then heat inactivate the restriction enzyme. At this stage the digested plasmid can be kept on ice or frozen until required. Once other components have been prepared (see methods below), and the transformation is ready to be performed, add the digested plasmid to the PCR plate in new wells (if space) or a new 96-well plate, ready for the yeast recombination.

3.2 Primer Design

1. Ensure accurate annotation of the BGC(s) of interest as far as is possible (*see* **Note 12** and reference [28]). Transcriptomics data can help with this task. Pay particular attention to accurate identification of start and stop codons of each gene to be cloned.
2. Design primers to amplify the gene(s) of interest. For efficient recombination to occur, a minimum of 30 bp homology (vector specific) is recommended. If using the pTYGS series of vectors described above, the following 30 base compound tails are required, dependent on which cassette you wish to place your gene(s) into, where "N_{20}" represents the unique gene-specific primer sequence (Table 1). Ensure the gene-specific primer sequence encompasses the predicted start codon (forward primer) and stop codon (reverse primer) of the gene. See Note 13 if using a plasmid other than the pTYGS series.
3. Order primers using a synthetic DNA ordering service at a final concentration of 10 μM, desalted in 96-well or 384-well plates.

3.3 PCR Amplification

Liquid transfers were carried out by the Freedom Evo Tecan liquid handling robot; however, feasibly any liquid handling robots capable of performing liquid transfers of a minimum volume of 1 μL can be used and should be programmed as per the manufacturer's instructions.

Table 1
Specific 30 base sequences required to provide homology to specific cassettes within the pTYGS series of vectors. "N_{20}" denotes where gene specific sequences should be placed within the primer

Cassette	Forward primer	Reverse primer
P_{amyB}	5′-tctgaacaataaaccccacagcaagctccgN_{20}-3′	5′-ctctccacccttcacgagctactacagatcN_{20}-3′
P_{adh}	5′-tctttcaacacaagatcccaaagtcaaaggN_{20}-3′	5′-cattctatgcgttatgaacatgttccctggN_{20}-3′
P_{gpdA}	5′-cagctaccccgcttgagcagacatcaccggN_{20}-3′	5′-gacaatgtccatatcatcaatcatgaccggN_{20}-3′
P_{eno}	5′-cgactgaccaattccgcagctcgtcaaaggN_{20}-3′	5′-ttggctggtagacgtcatataatcatacggN_{20}-3′

Table 2
Volume of reagents required for a 50 μL PCR reaction

Reagents	Volume (μL/reaction)
Sterile water	22
PCR master mix	25
Template DNA (10 ng/μL)	1
Forward primer (10 μM)	1
Reverse primer (10 μM)	1

1. PCR reactions should be set up using a liquid handling robot to ensure efficiency for high-throughput studies. Prepare a PCR master mix with a high-fidelity polymerase according to the manufacturer's protocol (*see* **Note 2**). Add the master mix, sterile water, and template DNA (~10 ng/μL) in appropriate positions within the liquid handling robot. Then position the primers (10 μM) ordered in 96- or 384-well plates. Program the robot to pipette each primer, template DNA, sterile water, and PCR master mix (volumes in Table 2) into each well on a 96- or 384-well plate, where each well represents a separate gene to be amplified.
2. Program the robot to move the PCR plate to the integrated thermocycler (*see* **Note 14** for our experience with robotic manipulation of plates). Alternatively, if this add-on is not present, manually transfer the PCR plate to an external thermocycler. Run the PCR program recommended in the polymerase protocol. Calculate the PCR extension time based on the largest gene to be amplified on a given plate. As a rule of thumb, an extension time of 1 min/kb can be used but this may differ based on the polymerase being used. Calculate the annealing temperature based on recommendations in the polymerase protocol (*see* **Note 15**).
3. Run 5 μL of all PCR reactions on an agarose gel and ensure all the PCRs have amplified correctly, reserving the rest for recombination. For those that may not have worked, repeat these manually and replace the amplified product in the correct position in the PCR plate.

3.4 Construction of the Multi-gene Expression Plasmids via Yeast Recombination

1. Streak out your yeast strain (*see* **Note 3** for details on yeast strain) on solid YPAD and incubate at 30 °C (*see* **Note 16**) until clear colonies appear (2–3 days).
2. Inoculate at least one 10 mL YPAD starter culture (*see* **Note 17** for details on amounts required) with a single colony of *S. cerevisiae* and grow overnight at 30 °C with shaking at 200 rpm.

3. Add one 10 mL overnight starter culture to 40 mL of YPAD in a 250 mL Erlenmeyer (conical) flask (scale number of flasks as required) and incubate at 30 °C with shaking at 200 rpm for 4–5 h, or until the culture reaches at least 2×10^7 cells mL^{-1} (*see* **Note 18**).
4. Centrifuge the culture(s) at 3000 × g for 5 min and discard the supernatant.
5. Wash each culture of cells with 25 mL sterile H_2O and repeat the above centrifugation, then resuspend each pellet in 1 mL 0.1 M LiOAc and transfer to a 1.5 mL microcentrifuge tube.
6. Pellet each set of cells at 16,000 × g for 15 s and discard the supernatant, then resuspend the cells in 400 μL 0.1 M LiOAc (*see* **Note 19**).
7. Combine the competent yeast cells in a single 15 mL centrifuge tube. Keep on ice until use and then transfer to the appropriate chilled rack in the robot.
8. Prepare the transformation plates by autoclaving sufficient appropriate drop-out medium (*see* **Note 20**) (e.g., SM-URA agar (0.68% (w/v) yeast nitrogen base, 0.5% (w/v) ammonium sulphate, 2% (w/v) glucose, 0.077% (w/v) complete supplement mixture minus uracil, 1.5% (w/v) agar)). Pipette 2 mL into each 24-well recovery plate (where each well represents a single transformation) and leave aside to set until the plate is required in step 11*v*.
9. Prepare the additional reagents in Table 3 required for transformation, ensuring the extra volume is also supplied where necessary, to compensate for the volume that remains inaccessible at the bottom of the trough/tube when pipetting. The pre-mixed and pre-denatured SS-DNA and 1 M LiOAc should be provided in 8 x 1 mL microcentrifuge tubes (1:1.4 ratio of SS-DNA in 2 mg/mL in TE buffer:1 M LiOAc). Ensure the tubes of SS-DNA and LiOAc are kept on ice after thawing (*see* **Notes 5** and **6** for details on preparation and storage).
10. Place the sterile water and sterile PEG in separate troughs in the robot. Place competent yeast cells and the 8 x mix of denatured SS-DNA and 1 M LiOAc in the chilled rack in the robot (*see* **Notes 5** and **6** for details on preparation and storage). Lastly, position the PCR product/digested plasmid plate and 96-well transformation plates, ready for automated recombination (*see* Fig. 3).
11. Program the robot as per the manufacturer's instructions to carry out the steps below (*see* Fig. 3). If using the Freedom Evo Tecan liquid handling robot, see script in the additional document.

Table 3
Volume of reagents required for each transformation. Number in square brackets indicates the extra volume to compensate for the volume that remains inaccessible at the bottom of the trough/tube when pipetting

Reagent for transformation	Volume per transformation (μL)	Total volume per 96-well plate [extra volume required] (mL)	
Sterile 50% w/v polyethylene (PEG) 3350	240	23.04 [1]	
Sterile water	300	28.8 [1]	
Denatured SS-DNA and 1 M LiOAc mix	86	8.26 [0.8] (in total)	[b]1.03 [0.1] (per tube)
PCR product(s)	~5[a]	n/a	
Digested plasmid	5	0.48 [0.02]	
Competent yeast cells	50	4.8 [0.1]	

[a]Approximately equal amounts of each DNA fragment required for the recombination should be added. Estimate concentration from the DNA gel, and adjust volume here accordingly.
[b]For our set-up it was necessary to split the denatured SS-DNA and 1 M LiOAc mix into 8 x Tubes to allow for the mix to be kept chilled

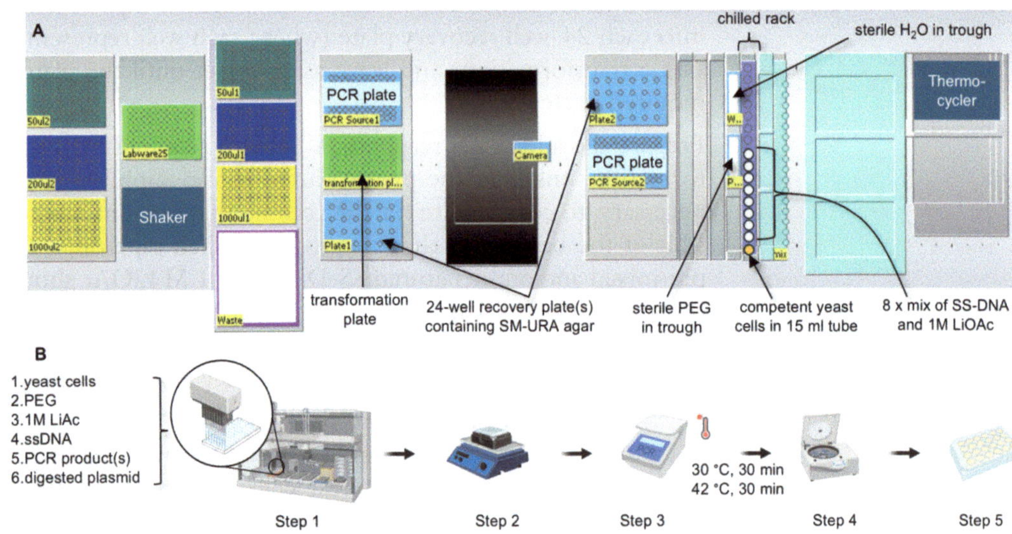

Fig. 3 **(a)** Birds eye view of a liquid handling robot, highlighting the setup of reagents and in-built equipment for PCR amplification and yeast recombination. **(b)** Workflow of the yeast recombination protocol. *Step 1:* Transformation reagents combined in a 96-well plate, where each well represents a single recombination. *Step 2:* Transformation reagents mixed by transfer of 96-well plate to integrated/external shaker. *Step 3*: Incubation of 96-well plate at 30 °C for 30 min followed by 42 °C for 30 min, by transfer to integrated/external thermocycler. *Step 4:* Centrifugation of cells to pellet cells and remove supernatant. *Step 5:* Plating of transformed cells on pre-prepared 24-well recovery plate

(i) Robot to transfer each transformation reagent (*see* Table 3 for volumes) to the 96-well transformation plate (*see* **Notes 21** and **22**).

(ii) Robot to transfer transformation plate (*see* **Note 14**) to integrated shaker to vigorously mix transformation reagents together (*see* **Note 23**).

(iii) Robot to transfer transformation plate (*see* **Note 14**) to integrated thermocycler and incubate cells at 30 °C for 30 min and then 42 °C for 30 min (or external thermocycler/heating block).

(iv) Manually move the transformation plate to a centrifuge, pellet the cells at 3000 × *g* for 15 s, and then invert the plate on a paper towel to carefully remove the supernatant.

(v) Place the transformation plate back in the robot and place the pre-prepared 24-well recovery plate(s) in the appropriate place(s) in the robot.

(vi) Robot to pipette 0.3 mL of sterile water into each well in the transformation plate and pipette mix to resuspend cells (*see* **Note 24**).

(vii) Robot to pipette 0.1 mL of transformed cell to the 24-well recovery plate, where each well represents a single recombination.

(viii) Manually transfer the 24-well recovery plate to an incubator and incubate cells at 30 °C for 3–4 days until colonies appear.

12. Once colonies appear continue with *E. coli* shuttling as per the protocol in ref. [27] (*see* **Note 25**).

Our current protocol does not include automated screening of *E. coli* colonies; however, this is envisaged as a future enhancement to further streamline the process.

4 Notes

1. We use a Freedom Evo Tecan liquid handling robot; however, other robots are available from manufacturers such as Hamilton and Beckman with similar capabilities.
2. We routinely use KAPA HiFi HotStart ReadyMix (Roche).
3. We routinely use *S. cerevisiae* strains YPH499 or YPH500.
4. To prepare 100 mL of 50% (w/v) PEG solution, first add 50 g of PEG 3350 to approximately 30 mL of deionized H_2O then stir or incubate at 50 °C until fully dissolved. Top up to 100 mL

with H_2O and autoclave. PEG must be stored in a securely capped bottle to prevent evaporation as this severely impacts transformation efficiency.

5. To prepare salmon sperm DNA (SS-DNA) for yeast recombination, dissolve 200 mg of high molecular weight salmon sperm DNA (e.g., Sigma-Aldrich catalogue number D1626) in 100 mL of sterile TE using a magnetic stirrer at 4 °C. This may take some hours. Pipetting the solution gently with a wide-bore pipette can speed up dissolution. Aliquots of SS-DNA should be stored at −20 °C.
6. Prior to use, the SS-DNA must be denatured by boiling in a water bath for 5 min, then chilled immediately on ice before use. A similar process can be achieved by programming a PCR thermocycler to 98 °C for 5 min prior to holding at 4 °C. Unused denatured SS-DNA/1 M LiOAc mix can be returned to the freezer and can be used immediately after thawing, as long as it has not been allowed to warm above 4 °C, without the need to denature again. It can go through 3–4 freeze-thaw cycles before further denaturing is required. If repeated denaturing is required, SS-DNA 1 M LiOAc mix can be boiled three or four times without significant loss of activity.
7. We use the ccdB resistant One Shot™ ccdB Survival™ 2 T1R Competent Cells available from Thermo Fisher Scientific.
8. All the pTYGS plasmids are available on request.
9. This depends on your BGC and the experiments you wish to undertake. For example, if you have a BGC with 8 genes which you wish to co-express, you should clone 4 genes into pTYGS-arg, and 4 genes into pTYGS-ade. For an example of this strategy, please see reference [36]. *See* **Note 10** for information on cloning order.
10. In general, we attempt to predict the order of biosynthesis using bioinformatics, to allow us to add the "early" genes to one plasmid and "later" genes to further plasmids. We use arginine prototrophy for selection preferentially, as it gives the least background growth. Adenine prototrophy is also reliable. Methionine prototrophy can be less reliable, and it is important to check that there are no sources of methionine in your recovery medium; for example, we have found that some batches of agar are impure and will not support selection for methionine.
11. In most cases, digestion using both *NotI* and *SgsI* will be necessary, unless you wish to leave the amyB cassette empty (*NotI* digest not required), or all three *SgsI* cassettes (adh, gpdA, eno) empty (*SgsI* not required). We conduct large-scale digests overnight to ensure complete digestion.

12. We routinely use AUGUSTUS [37] and fgenesh [38], coupled with manual annotation.
13. If using a different plasmid ensure the primers are flanked with desired homology to promoters, terminators, adjacent DNA or the vector backbone. In theory, you can start with an empty yeast vector with a selectable marker and can recombine in genes with promoters and terminators by including them in the primer sequence.
14. We have found that the robotic manipulator arm for our Freedom Evo Tecan liquid handling robot to be unreliable. Performing a "warm-up" sequence led to fewer accidents; however, to ensure the smooth progression of our experiments, we manually transferred plates throughout the protocol rather than relying on the manipulator arm.
15. We use 60 °C as standard annealing temperature when using KAPA HiFi HotStart ReadyMix (Roche).
16. Although 30 °C is the standard temperature for culturing *S. cerevisiae*, we routinely incubate at 28 °C for convenience (also applies to steps 2, 3, and 11viii within Subheading 3.4). This does not appear to hinder the yeast recombination in any way, although growth rates may be slightly slower.
17. Up to 6 mL of competent cells can be prepared depending on how many transformations are planned. 50 μL is required per transformation, and controls must be included (ideally a sample of each digested vector used as negative controls, and uncut plasmid as a positive control). Final resuspension of the cells will produce approx. double the volume of 0.1 M LiOAc used (e.g., if resuspended in 400 μL, the final volume of cells is approx. 800 μL). Ensure at least 100 μL extra cells are included to account for pipetting error.
18. For strain YPH499, we find a 4-h incubation reliably results in an efficient transformation, but if a different yeast strain is being used, or transformation efficiencies are low, quantifying the yeast cells at this point may be worthwhile. This can be done using a hemocytometer or spectrophotometer. An OD_{600} of approximately 1.5 should correspond with a cell concentration of 2×10^7 cells mL^{-1}, but as OD_{600} can vary between different yeast strains and spectrophotometers, we would recommend performing an initial calibration by comparing the OD_{600} with cell counts using a hemocytometer.
19. Cells at this stage can be stored on ice for up to 4 weeks prior to use, although a reduction in efficiency may be observed. Alternatively, cells can be prepared for freezing and long-term storage, according to the protocol published by Gietz et al. [39]

20. Each well in the 24-well plate will contain medium for the growth of yeast cells in which a single specific set of recombination events has occurred. Prepare 2 mL x number of recombined plasmids expected, plus ~5 mL extra of the appropriate yeast drop-out medium.
21. Due to the viscosity of PEG, the PEG transfer must be programmed with care to ensure accurate pipetting. This step is optimized within our script.
22. In the original protocol, a total of 34 μL of DNA/sterile H_2O is added to the transformation. Here, only the required DNA fragments are added. No extra sterile H_2O is added.
23. If no integrated shaker is present, manually mix the transformation components using a plate vortexer.
24. The original protocol uses 1 mL H_2O to resuspend yeast cells [27], however testing has shown that 0.3 mL is sufficient, and allows for resuspension within the 96-well plate.
25. Due to the small size of the well, it is likely that no single colonies will be identifiable. With our protocol outlined in ref. [27] all colonies on the yeast transformation plate are simultaneously utilized for plasmid extraction. Screening for correct recombination events then occurs at the *E. coli* stage.

Acknowledgments

We would like to thank MRC (MR/N029909/1) and UWE (New Starter Research Development Scheme) for funding for KW, and the SWBio PhD studentship (BB/T008741/1) funding for ADS. We would like to thank BrisSynBio for the use of the Freedom Evo Tecan, which was funded by BBSRC and EPSRC (BB/L01386X/1). We would also like to thank Dr. Peter Wilson for help with programming the Freedom Evo Tecan. Figure 1 was created in BioRender. Williams, K. (2025) https://BioRender.com/r60v422. Figure 3 was created in BioRender. Williams, K. (2025) https://BioRender.com/t471111.

References

1. Frisvad JC, Samson RA (2004) Polyphasic taxonomy of *Penicillium* subgenus *Penicillium* A guide to identification of food and air-borne terverticillate Penicillia and their mycotoxins. Stud Mycol 49:1–174
2. Grove JF (1963) Griseofulvin. Q Rev Chem Soc 17:1–19
3. Ruegger A, Kuhn M, Lichti H et al (1976) Cyclosporin-A, a peptide metabolite from *Trichoderma polysporum* (LINK ex PERS.) *Rifai*, with a remarkable immunosuppressive activity. Helv Chim Acta 59:1075–1092
4. Bailey AM, Alberti F, Kilaru S et al (2016) Identification and manipulation of the pleuromutilin gene cluster from *Clitopilus passeckerianus* for increased rapid antibiotic production. Sci Rep 6:1–11
5. Knauseder F, Brandl E (1976) Pleuromutilins. Fermentation, structure and biosynthesis. J Antibiot 29:125–131

6. Mulder KCL, Mulinari F, Franco OL et al (2015) Lovastatin production: from molecular basis to industrial process optimization. Biotechnol Adv 33:648–665
7. Manzoni M, Rollini N (2002) Biosynthesis and biotechnological production of statins by filamentous fungi and application of these cholesterol-lowering drugs. Appl Microbiol Biotechnol 58:555–564
8. Schramm G, Steglich W, Anke T, Oberwinkler F (1978) Antibiotics from basidiomycetes, III. Strobilurin A and B, antifungal metabolites from *Strobilurus tenacellus*. Chem Berichte-Recueil 111:2779–2784
9. Nofiani R, de Mattos-Shipley KMJ, Lebe KE et al (2018) Strobilurin biosynthesis in basidiomycete fungi. Nat Commun 9:1–11
10. Wu B, Hussain M, Zhang W et al (2019) Current insights into fungal species diversity and perspective on naming the environmental DNA sequences of fungi. Mycology 10:127–140
11. Prescott TAK, Hill R, Mas-Claret E et al (2023) Fungal drug discovery for chronic disease: history, new discoveries and new approaches. Biomolecules 13:1–25
12. Blin K, Shaw S, Augustijn HE et al (2023) antiSMASH 7.0: new and improved predictions for detection, regulation, chemical structures and visualisation. Nucleic Acids Res 51: W46–W50
13. Zdouc MM, Blin K, Louwen NLL et al (2025) MIBiG 4.0: advancing biosynthetic gene cluster curation through global collaboration. Nucleic Acids Res 53:D678–D690
14. Hannigan GD, Prihoda D, Palicka A et al (2019) A deep learning genome-mining strategy for biosynthetic gene cluster prediction. Nucleic Acids Res 47:e110–e110
15. Skinnider MA, Dejong CA, Rees PN et al (2015) Genomes to natural products PRediction informatics for secondary metabolomes (PRISM). Nucleic Acids Res 43:9645–9662
16. Gilchrist CLM, Booth TJ, van Wersch B et al (2021) Cblaster: a remote search tool for rapid identification and visualization of homologous gene clusters. Bioinform Adv 1:1–10
17. Umemura M, Koike H, Nagano N et al (2013) MIDDAS-M: motif-independent de novo detection of secondary metabolite gene clusters through the integration of genome sequencing and transcriptome data. PLoS One 8:e84028
18. Greco C, Keller NP, Rokas A (2019) Unearthing fungal chemodiversity and prospects for drug discovery. Curr Opin Microbiol 51:22–29
19. Cacho RA, Tang Y, Chooi Y-H (2015) Next-generation sequencing approach for connecting secondary metabolites to biosynthetic gene clusters in fungi. Front Microbiol 5:1–16
20. Barbesgaard P, Heldthansen HP, Diderichsen B (1992) On the safety of *Aspergillus oryzae* – a review. Appl Microbiol Biotechnol 36:569–572
21. Skellam E, Rajendran S, Li L (2024) Combinatorial biosynthesis for the engineering of novel fungal natural products. Commun Chem 7:89
22. Wiemann P, Soukup AA, Folz JS et al (2018) CoIN: co-inducible nitrate expression system for secondary metabolites in *Aspergillus nidulans*. Fungal Biol Biotechnol 5:1–10
23. Clevenger KD, Bok JW, Ye R et al (2017) A scalable platform to identify fungal secondary metabolites and their gene clusters. Nat Chem Biol 13:895–901
24. Nielsen MR, Sørensen JL (2022) Investigating fungal biosynthetic pathways using heterologous gene expression: *Fusarium* sp. as a heterologous host. In: Skellam E (ed) Methods in molecular biology. Humana Press Inc.
25. Jin FH, Maruyama J-I, Juvvadi PR et al (2004) Development of a novel quadruple auxotrophic host transformation system by *argB* gene disruption using *adeA* gene and exploiting adenine auxotrophy in *Aspergillus oryzae*. FEMS Microbiol Lett 239:79–85
26. Lazarus CM, Williams K, Bailey AM (2014) Reconstructing fungal natural product biosynthetic pathways. Nat Prod Rep 31:1339–1347
27. de Mattos-Shipley KMJ, Lazarus CM, Williams K (2022) Investigating fungal biosynthetic pathways using heterologous gene expression: *Aspergillus oryzae* as a heterologous host. In: Skellam E (ed) Engineering natural product biosynthesis: methods and protocols. Humana Press Inc., New York
28. Navarro-Muñoz JC, Collemare J (2022) A bioinformatics workflow for investigating fungal biosynthetic gene clusters. Methods Mol Biol 2489:1–21
29. Gomi K, Iimura Y, Hara S (1987) Integrative transformation of *Aspergillus oryzae* with a plasmid containing the *Aspergillus nidulans argB* gene. Agric Biol Chem 51:2549–2555
30. Yamada O, Lee BR, Gomi K (1997) Transformation system for *Aspergillus oryzae* with double auxotrophic mutations, *niaD* and *sC*. Biosci Biotechnol Biochem 61:1367–1369
31. Tada S, Gomi K, Kitamoto K et al (1991) Construction of a fusion gene comprising the Taka-amylase-A promoter and the *Escherichia coli* beta-glucuronidase gene and analysis of its

expression in *Aspergillus oryzae*. Mol Gen Genet 229:301–306

32. Pahirulzaman KAK, Williams K, Lazarus CM (2012) A toolkit for heterologous expression of metabolic pathways in *Aspergillus oryzae*. In: Hopwood DA (ed) Methods in enzymology. Academic.
33. Zakian VA, Brewer BJ, Fangman WL (1979) Replication of each copy of the yeast 2 micron DNA plasmid occurs during the S phase. Cell 17:923–934
34. Gerbaud C, Fournier P, Blanc H et al (1979) High frequency of yeast transformation by plasmids carrying part of entire 2-μm yeast plasmid. Gene 5:233–253
35. Raymond CK, Pownder TA, Sexson SL (1999) General method for plasmid construction using homologous recombination. BioTechniques 26:134–141
36. Williams K, Szwalbe AJ, Mulholland NP et al (2016) Heterologous production of fungal maleidrides reveals the cryptic cyclization involved in their biosynthesis. Angew Chem Int Ed 55:6784–6788
37. Stanke M, Morgenstern B (2005) AUGUSTUS: a web server for gene prediction in eukaryotes that allows user-defined constraints. Nucleic Acids Res 33:W465–W467
38. Solovyev V, Kosarev P, Seledsov I, Vorobyev D (2006) Automatic annotation of eukaryotic genes, pseudogenes and promoters. Genome Biol 7:S10
39. Gietz RD, Schiestl RH (2007) Frozen competent yeast cells that can be transformed with high efficiency using the LiAc/SS carrier DNA/PEG method. Nat Protoc 2:1–4

Chapter 8

Molecular Lego in the Green Factory: Engineering Plant Metabolism with Golden Gate Assembly

Rajesh Chandra Misra, Ramesha Thimmappa, and Anne Osbourn

Abstract

Plant specialized metabolites encompass a diverse array of bioactive compounds with significant pharmaceutical, nutraceutical, and industrial relevance, and they play vital roles in plant defense, signaling, and ecological interactions. However, their complex biosynthetic pathways and low natural yields present a major bottleneck for commercial exploitation. Synthetic biology provides new tools for the rational design and engineering of metabolic pathways to enhance specialized metabolite biosynthesis in plants as well as in microbes such as *E. coli* and yeast. Here, we report a synthetic biology framework utilizing Golden Gate assembly (GGA) for the modular construction and optimization of plant metabolic pathways. Golden Gate enables scarless, directional assembly of multigene constructs with high efficiency, ideal for refactoring biosynthetic gene clusters and for the systematic design of complex metabolic networks. Golden Gate assembly can efficiently join up to 10–20 fragments simultaneously in a single reaction, a significant improvement over traditional molecular biology cloning methods. We developed and validated a standardized module of promoters, untranslated regions (UTRs), coding sequences, and regulatory elements compatible with plant transformation systems. Using this platform, we successfully reconstructed the biosynthetic pathway for plant specialized metabolites by transient expression in *Nicotiana benthamiana*, demonstrating significant improvements in construct stability. Our results highlight the potential of Golden Gate as a cornerstone technology for plant metabolic engineering, offering a scalable and versatile strategy to enhance the production of valuable specialized metabolites and set the stage for custom metabolic rewiring aimed at sustainable production of high-value compounds.

Key words Plant specialized metabolism, Golden Gate assembly, Metabolites, *Nicotiana benthamiana*, Biosynthesis, DNA fragments, Metabolic engineering, Type IIS enzymes

1 Introduction

The advent of genetic engineering has revolutionized plant science, enabling the precise manipulation of plant genomes for various applications, from enhancing crop yield and nutritional value to improving disease resistance. At the heart of this revolution lies the ability to efficiently assemble DNA constructs, which are the molecular blueprints for introducing new genetic information into plants [1, 2]. Traditional cloning methods, often relying on restriction

Hans-Wilhelm Nützmann and Ramesha Thimmappa (eds.), *Metabolic Gene Clusters: Methods and Protocols*, Methods in Molecular Biology, vol. 3054, https://doi.org/10.1007/978-1-0716-5384-5_8,

enzymes and DNA ligase, can be cumbersome, time-consuming, and limited by the availability of unique restriction sites. The need for more efficient, standardized, and high-throughput cloning strategies became apparent with the increasing complexity of genetic engineering projects, particularly in the context of multigene constructs and synthetic biology [3, 4]. The field of synthetic biology is progressing at a rapid pace, driven by significant advances in various interconnected scientific and technological areas. These include genetic engineering, molecular biology, DNA synthesis, and computational design. Recent developments include the engineering of organisms with entirely synthetic genomes, through the design and synthesis of new genetic material that can be integrated into living cells [5, 6]. These synthetic genomes have the potential to revolutionize various applications, such as producing pharmaceuticals, biofuels, and environmentally friendly materials, as well as enabling the development of novel organisms with tailored functions [7]. While significant advancements have been made in DNA manipulation, the available toolkit for assembling DNA fragments remained quite limited.

Golden Gate assembly, a powerful Type IIS restriction enzyme-based cloning method, has emerged as a transformative tool in this landscape. Its ability to seamlessly assemble multiple DNA fragments in a single, one-pot reaction, without leaving unwanted "scar" sequences, makes it exceptionally well-suited for the modular construction of complex genetic circuits [8–10]. Unlike Type IIP enzymes, which cut within their recognition sequence, Type IIS enzymes such as *Bsa*I, *Bpi*I (*Bbs*I), and *Sap*I cut at a defined distance outside their recognition sequence. Each of these enzymes recognizes a specific, typically asymmetric, recognition site (e.g., *Bsa*I recognizes GGTCTC and B*pi*I recognizes GAAGAC) [11]. After binding to this site, the enzyme cleaves the DNA a few base pairs away, generating a short, single-stranded overhang (typically 4 base pairs). The entire assembly process typically occurs in a single reaction vessel and one of the most critical advantages is the scarless nature of the assembly [12, 13]. The crucial aspect is that the recognition site itself is removed during the cleavage, meaning it is not present in the assembled product. By carefully designing the DNA fragments (often referred to as "parts" or "modules") to have complementary overhangs at their ends, multiple fragments can be joined together in a precise, directional manner and crucial for maintaining the precise reading frames of coding sequences and ensuring the correct functionality of regulatory elements. For example, if fragment X has an overhang AAAA at its 3′ end and fragment Y has a complementary overhang TTTT at its 5′ end, they can be ligated together. In plant transformation, where precise gene expression is often paramount, scarless assembly minimizes potential artifacts. Additionally, Golden Gate assembly can be performed hierarchically, allowing for the construction of increasingly complex genetic constructs. For instance, individual gene components (promoters, coding sequences, terminators) can be assembled into "transcriptional units" in a level-1 assembly. These transcriptional units can then be combined with other level-1 constructs and backbone vectors in a level-2 assembly to create multi-gene constructs [14]. Moreover, the modular cloning system (MoClo) allows for the rapid construction of complex genetic constructs using a defined set of pre-made vectors and a defined assembly strategy. This modularity greatly simplifies the design and construction of large genetic circuits. Libraries of standardized DNA "parts" (e.g., promoters with specific activities, various coding sequences, different terminators), each flanked by universal Type IIs recognition sites and unique overhangs, can be created [9]. These parts can then be easily interchanged and assembled in various combinations, accelerating the design and testing of new genetic constructs. This standardization facilitates collaboration and sharing of genetic resources within the plant science community.

The one-pot, one-step nature of the reaction, combined with the continuous cycling of cutting and ligating, leads to exceptionally high assembly efficiency. This makes it ideal for high-throughput cloning applications, allowing researchers to rapidly generate numerous constructs for screening or functional studies. The modularity of Golden Gate assembly is a cornerstone of plant synthetic biology, where many different combinations of novel genetic circuits, including biosensors for detecting environmental stimuli or internal plant states, can be designed, assembled, and tested [15]. The most fundamental application is the rapid and efficient construction of gene expression cassettes and enabling the assembly of several transcriptional units into a single T-DNA vector for *Agrobacterium*-mediated transformation [16, 17]. This has been instrumental in engineering plants for enhanced nutrient biosynthesis (e.g., Golden Rice for Vitamin A), increased biomass production, or the synthesis of valuable secondary metabolites. Additionally, Golden Gate assembly is widely used for constructing CRISPR-Cas gene editing vectors. Guide RNA (gRNA) expression cassettes, *Cas* enzyme expression cassettes, and donor DNA templates for homologous recombination can all be efficiently assembled into a single vector [18]. This simplifies the creation of tools for precise genome editing in plants, allowing for targeted gene knockouts, insertions, or base changes. Further, integration with robotic platforms will further increase throughput, enabling the rapid construction of thousands of plant expression constructs for large-scale functional genomics, synthetic biology, and breeding programs [19]. While Golden Gate assembly offers numerous advantages, there are some considerations and potential challenges. The initial design of DNA parts requires careful planning to avoid unintended assemblies or self-ligation. While the overall time and effort savings are substantial, the initial cost of Type IIS restriction enzymes and specialized kits can be higher than basic restriction enzymes. Other methods like Gibson assembly and GoldenBraid offer seamless cloning without restriction site constraints and slightly more flexible in dealing with internal restriction sites due to its modular design, respectively, but may involve different costs and procedural complexities such as requires longer homologous overlaps, less versatile for multi-fragment assembly and complex workflow with multiple steps compared to the one-pot Golden Gate method [7, 9, 20, 21]. We provide here a detailed protocol for designing and modifying DNA parts including promoter, untranslated regions (UTRs), terminators, and coding sequences. The protocol also provides guidance on how to perform multigene assembly using the modular cloning system MoClo with the set of identified and/or designed parts and vectors.

2 Materials

2.1 Plant Material

1. Tobacco (*Nicotiana benthamiana*) plants were employed for transient expression assays due to their amenability to agroinfiltration.
2. *N. benthamiana* plants were grown in a controlled environment chamber at 24 °C under a 16 h light/8 h dark cycle.

2.2 Bacterial Strains

1. *E. coli DH5α*: For cloning and plasmid propagation and was maintained at 37 °C.
2. *Agrobacterium tumefaciens* strain GV3101: For plant agrofiltration (transient).

2.3 Bacterial Growth Media and Antibiotics

1. Luria-Bertani broth: Tryptone (10 g/L), Yeast extract (5 g/L), NaCl (10 g/L). Combine reagents and dissolve in 800 mL deionized water, adjust pH to 7.0 using NaOH (1 M), and finally adjust volume to 1 L.
2. Luria-Bertani agar: Tryptone (10 g/L), Yeast extract (5 g/L), NaCl (10 g/L). Combine reagents and dissolve into 800 mL deionized water and adjust pH to 7 using NaOH (1 M). Add 20 g of agar and finally adjust volume to 1 L.
3. 50 mg/mL spectinomycin, carbenicillin, kanamycin filter-sterilized stock solution in sterilized Milli-Q water and 30 mg/mL rifamycin in dimethylformamide (DMF).

2.4 DNA Components

1. PCR-amplified or synthesized DNA fragments flanked by complementary Type IIS restriction sites that generate specific 4-base overhangs for directional assembly.

2. Destination vector (acceptor plasmid) designed with appropriate Type IIS restriction sites and a selectable marker (e.g., antibiotic resistance gene, lacZ for blue/white screening) or a plant-compatible binary vector with a T-DNA region for *Agrobacterium*-mediated transformation.

2.5 Enzymes and Reagents

1. Type IIS restriction enzymes: 10 U/μL of *Bsa*I, *Bpi*I (*Bbs*I), *Esp*31 (Thermo Fisher). The specific enzyme used will depend on the design of the Golden Gate compatible DNA parts.
2. T4 DNA Ligase (400 U/μL, New England Biolabs).
3. T4 DNA Ligase Buffer (10X) containing ATP (supplied with T4 DNA ligase).
4. Bovine serum albumin (BSA, 10X).
5. 5-bromo-4-chloro-3-indolyl-β-D-galactopyranoside (X-gal): stock solution of 20 mg/mL in DMF (dimethylformamide).
6. Isopropyl β-D-1-thiogalactopyranoside (IPTG): stock solution 100 mg/mL in water.
7. 2-Morpholinoethanesulphonic acid (1 M, pH 5.6).
8. Magnesium chloride ($MgCl_2$, 1 M).
9. Acetosyringone (200 mM) in DMF.
10. Nuclease-free water.

2.6 PCR and Gel Electrophoresis

1. High-fidelity DNA polymerase (Phusion or Q5).
2. Deoxynucleoside triphosphate (dNTPs): 10 mM each.
3. PCR primers including Type IIS recognition sites and compatible overhangs.
4. Agarose powder (Sigma) for agarose gel electrophoresis (Agarose gel 1%): 1.0 g of agarose in 100 mL 1X TAE buffer.
5. GeneRuler™ 1 kb Plus DNA ladder (Thermo Fisher).
6. Tris-acetate-EDTA (TAE, 50X) buffer stock solution: Tris-HCl (2 M, pH 8.0), glacial acetic acid (1 M), EDTA (50 mM).
7. Gel extraction and PCR cleanup kits (Qiagen) for purifying DNA fragments.

2.7 General Laboratory Equipment, Consumables, and Software

1. Thermocycler and agarose gel electrophoresis unit.
2. Refrigerated centrifuge.
3. Incubator shaker for 28 °C and 37 °C.
4. Bio-Rad Gene Pulser Xcell.
5. Sterile 0.2 mL PCR tubes and 1.5 mL/2 mL microcentrifuge tubes.
6. Micropipettes and sterile tips.
7. 1 mL syringe (needleless).

8. Spectrophotometer for measuring DNA concentration.
9. UV transilluminator.
10. Gel analysis software.
11. DNA analysis software for generation of plasmid maps, such as Geneious Prime, SnapGene or A Plasmid Editor (ApE) software.

3 Methods

3.1 DNA Constructs

To advance the resource of plant toolkit for Golden Gate cloning and make them more versatile, new building blocks were designed based on plant binary vector pEAQ-*HT*-DEST1 [17]. We used this vector as a reference and divided its T-DNA region between right border (RB) and left border (LB) into seven parts (Fig. 1a).

1. In silico sequence analysis revealed the presence of multiple internal *Bpi*I and *Bsa*I recognition sites. Undesirable cleavage of entry clone can be removed through silent mutations with overlapping primers (*see* **Note 1**).
2. To eliminate these restriction sites, a set of PCR primers bearing point mutation with the compatible overlapping four base pair (bp) overhangs were used to amplified amplicons ranging in size from 150 bp to 1300 bp that allowed simultaneous assembly of multiple DNA fragments in a one-reaction step at predesigned fusion sites (*see* **Note 2**).
3. A plasmid containing desired DNA sequences was used as a template for the PCR (*see* **Note 3**). For instance, the plant binary vector pEAQ-*HT*-DEST1 was used as a template to amplify non-coding sequence, whereas available pDONR clones were used as a template to amplify coding sequences.
4. The PCR amplicons were then run on an agarose gel and purified by gel extraction before being used in the Golden Gate reaction to remove unwanted DNA fragments and ensure high-quality inputs for the assembly (*see* **Note 4**).
5. To minimize errors in assembly, these domesticated DNA fragments were initially assembled into basic Golden Gate level-0 destination vectors (Fig. 1b).
6. The assembly was performed using a type IIS enzyme, i.e., *Bpi*I and T4 DNA ligase following standard restriction/ligation protocol. Each level-0 destination vector is flanked by two *Bsa*I restriction sites (GGTCTC and GAGACC at the 5′ and 3′ sites, respectively) with specified four base pair (bp) overhangs to facilitate the assembly of multiple parts into level-1 architecture. In contrast to level-1 destination vectors, the *lacZ* cassette in level-0 destination vectors are

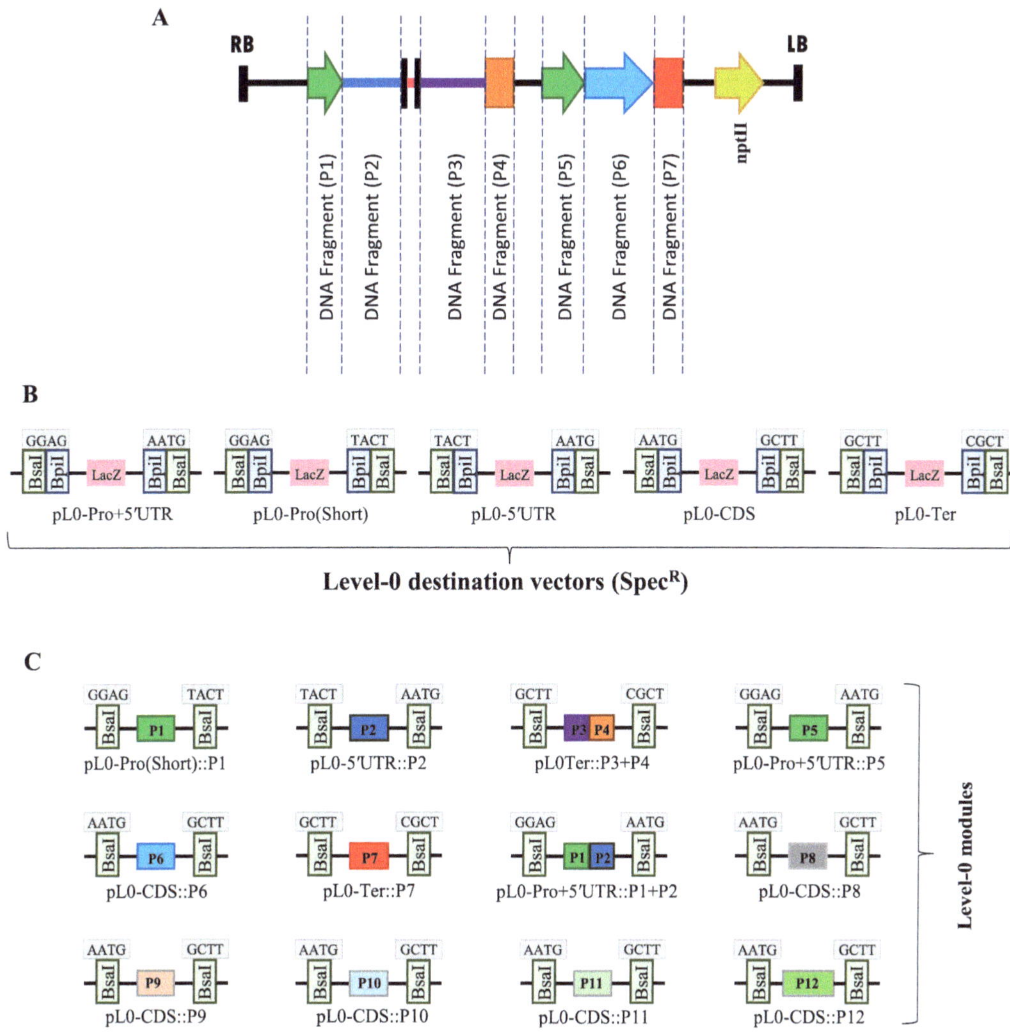

Fig. 1 Schematic representation of Golden Gate assembly strategy. (**a**) Plant gateway vector was used to build new level-0 parts for Golden Gate cloning. The T-DNA region between right border (RB) and left border (LB) was divided into seven parts, that represents P1, CaMV35s promoter (short); P2, 5'UTR Cowpea mosaic virus; P3, 3'UTR Cowpea mosaic virus P4, NOS terminator; P5, CaMV35s promoter (long); P6, Suppressor of silencing (P19) and P7, CaMV35s terminator. (**b**) Golden Gate level-0 destination vectors flanking with two *Bsa*I and *Bpi*I restriction sites and unique four-bp fusion sites are shown. These vectors encode a *lacZ* fragment for blue/white selection. (**c**) Golden Gate level-0 modules flanked by compatible fusion sites. P8 represents β-amyrin synthase, P9 represents cytochrome P450 C-28 oxidase, P10 represents cytochrome P450 C-16 oxidase, P11 represents cytochrome P450 C-23 oxidase, and P12 represents the green fluorescent protein (GFP). DNA fragments such as promoters, untranslated regions (UTRs), coding sequences, and terminators were cloned into level-0 destination vectors using a one-pot one-step cloning reaction via *Bpi*I that results in the release of cloned fragment(s) with *Bsa*I sites at both ends and the unique fusion sites identical to the used level-0 destination vector

flanked by two *Bpi*I and *Bsa*I recognition sites with non-identical fusion sites, which enable directional cloning.

7. The DNA fragments, i.e., P1, P2, P3, P4, P5, P6, P7, were cloned into respective level-0 destination vectors using *Bpi*I (Fig. 1c). As a result, DNA fragments were flanked by *Bsa*I sites with unique fusion sites, such as P1 was flanked by GGAG/TACT, P2 was flanked by TACT/AATG, P3 + P4 was flanked by GCTT/CGCT, P5 and P1 + P2 were flanked by GGAG/AATG, P7 was flanked by GCTT/CGCT, and P6, P8-P12 were flanked by AATG/GCTT (Fig. 1c) (*see* **Note 5**).
8. All level-0 modules (basic parts) were confirmed by sanger sequencing (*see* **Note 6**).

3.2 Assembly of Transcription Units

The placement and direction of gene of interest in the final construct determine which level-1 destination vector to use for assembling the transcription unit. A set of 14 vectors (pL1-[F1-F7] and pL1-[R1-R7]) are available to facilitate cloning in either forward or reverse orientation, respectively, at each of the seven possible positions (https://www.addgene.org/) [10].

1. We employed level-1 destination vectors aligned in the forward direction, containing GGAG and CGCT fusion sites to ensure compatibility with level-0 modules (Fig. 2a).
2. All level-1 destination vectors shared the same internal *Bsa*I sites, while external *Bpi*I sites enable compatibility across various positions. Interestingly, the downstream fusion site (TGCC) at position 7 of the level-1 destination vector is compatible with the upstream fusion site (TGCC) at position 1 of the same vector, allowing the assembly of over seven genes without requiring extra level-1 destination vectors (Fig. 2a). The unique four base pair fusion sites are compatible across level-1 destination vectors, allowing multiple level-1 modules to be directionally cloned into a level-2 destination vector [22].
3. After designing level-0 DNA modules—comprising individual genetic parts such as promoters, coding sequences, terminators, and others—assembling the transcription unit into the selected level-1 destination vector is straightforward, which involves pipetting all components into a reaction mixture and incubating it in a thermal cycler.
4. The assembly of three level-0 basic parts and the level-1 destination vector was performed via *Bsa*I following standard restriction/ligation protocol (Fig. 2b). The level-1 destination vector's cloning site includes a *lacZ* fragment that replaced during Golden Gate cloning in the MoClo system, enabling blue-white colony screening on selective media (Fig. 2b) (*see* **Note 7**) [9].

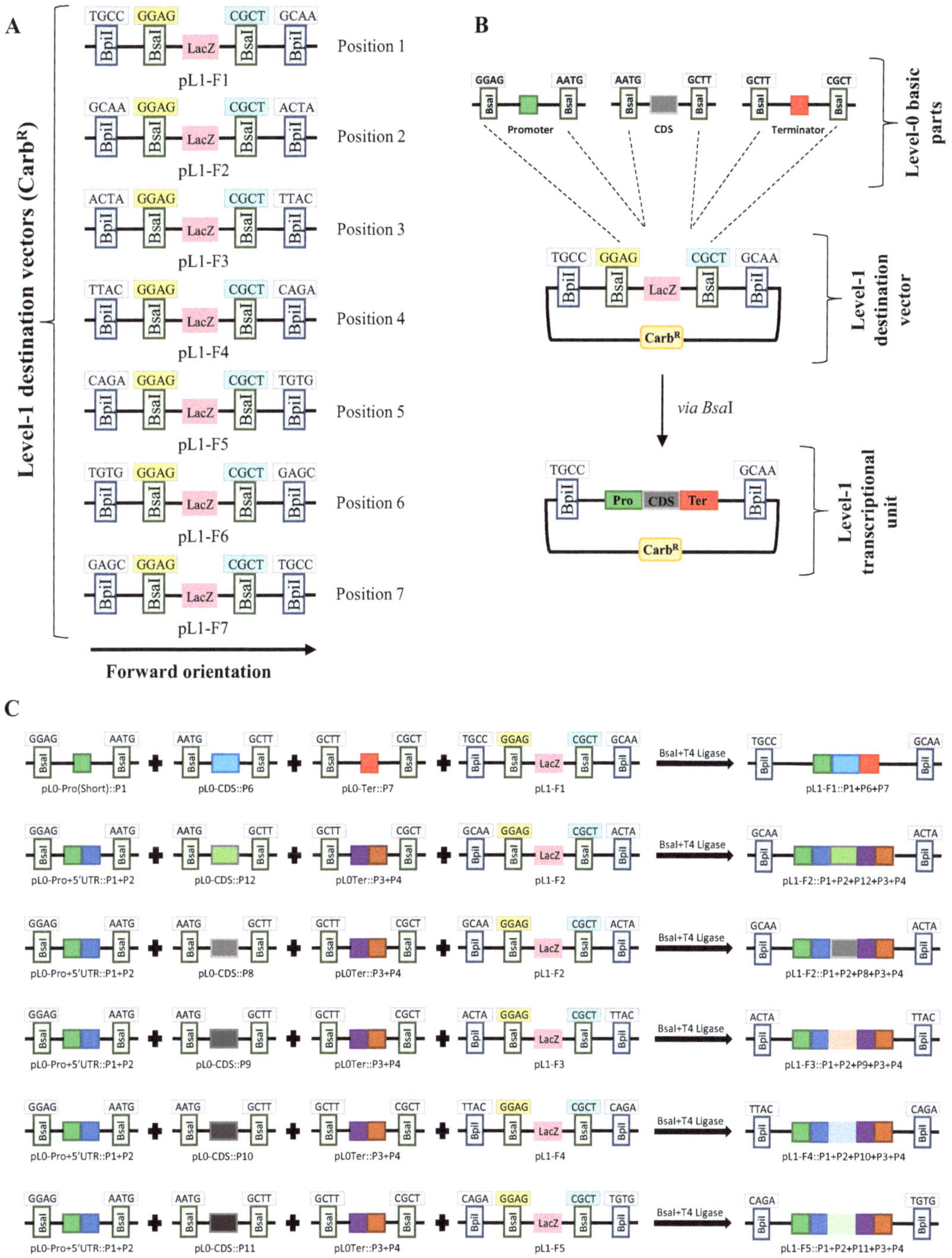

Fig. 2 Arrangement of type IIS restriction sites and fusion sites into universal level-1 destination vectors. (**a**) Level-1 destination vectors. (**b**) Assembly of a transcription unit consisting of three basic parts. (**c**) Strategically assembling various coding sequences (P6, P8, P9, P10, P11, P12) with compatible promoters and terminators into level-1 destination vectors at specific positions via *Bsa*I and T4 DNA ligase following standard restriction/ligation protocol

5. The metabolic pathway engineering was achieved by strategically assembling various coding sequences (P6, P8, P9, P10, P11, P12) with compatible promoters and terminators into level-1 destination vectors at specific positions (Fig. 2c).
6. To confirm the correct assembly of the plasmid, a restriction digestion was conducted using *Bpi*I, as the assembled transcription units were flanked by *Bpi*I recognition sites followed by sanger sequencing (*see* **Note 6**) (Fig. 2b, c).

3.3 Assembly of Multigene Construct

In a multi-level modular cloning system, multiple level-1 modules, which already contain individual DNA parts (like promoters or coding sequences), are joined together to form a larger, more complex construct within a level-2 destination vector. A collection of level-2 destination vectors is available that offers greater flexibility in designing multigene level-2 constructs. The level-2-destination vectors feature two *Bpi*I restriction sites flanking a DNA segment that encodes a red fluorescent protein (CRed), a color-selectable marker which allows for the selection of correctly assembled DNA constructs by visually distinguishing them from the original (Fig. 3a) [22]. Moreover, the vector backbone itself also contains an antibiotic resistance gene, i.e., kanamycin.

1. To enable precise, hierarchical assembly of multiple genetic parts each level-2 destination vector contains standardized upstream fusion sites such as TGCC, GCAA, ACTA, TTAC, CAGA, TGTG, GAGC that correspond to the ends of the level-1 modules, however the downstream fusion site (GGGA) is unique to level-2 destination vectors (Fig. 3a).
2. The GGGA fusion site is linked to the end of a previously assembled transcription unit using end-linkers. This molecular bridge provides the necessary *Bpi*I restriction sites to allow for a new round of Golden Gate assembly with a level-2 destination vector (Fig. 3b). However, this allows for the assembly of up to six level-1 modules, or transcriptional units, into a single level-2 construct as the *Bpi*I sites are eliminated from the final construct [18].
3. Here, we performed assembly of three level-1 modules (pL1-F1, pL1-F2, and pL1-F3), the appropriate end-linker (pEL3), and the level-2 destination vector (pL2-1) in a single one-pot, one-step cloning reaction in the presence of T4 DNA ligase and *Bpi*I (Fig. 3c). The level-2 destination vector's antibiotic resistance and CRed selection marker enable efficient counter-selection against level-1 module plasmids. Additionally, the red/white colorimetric system allows for easy identification of correctly assembled level-2 constructs.

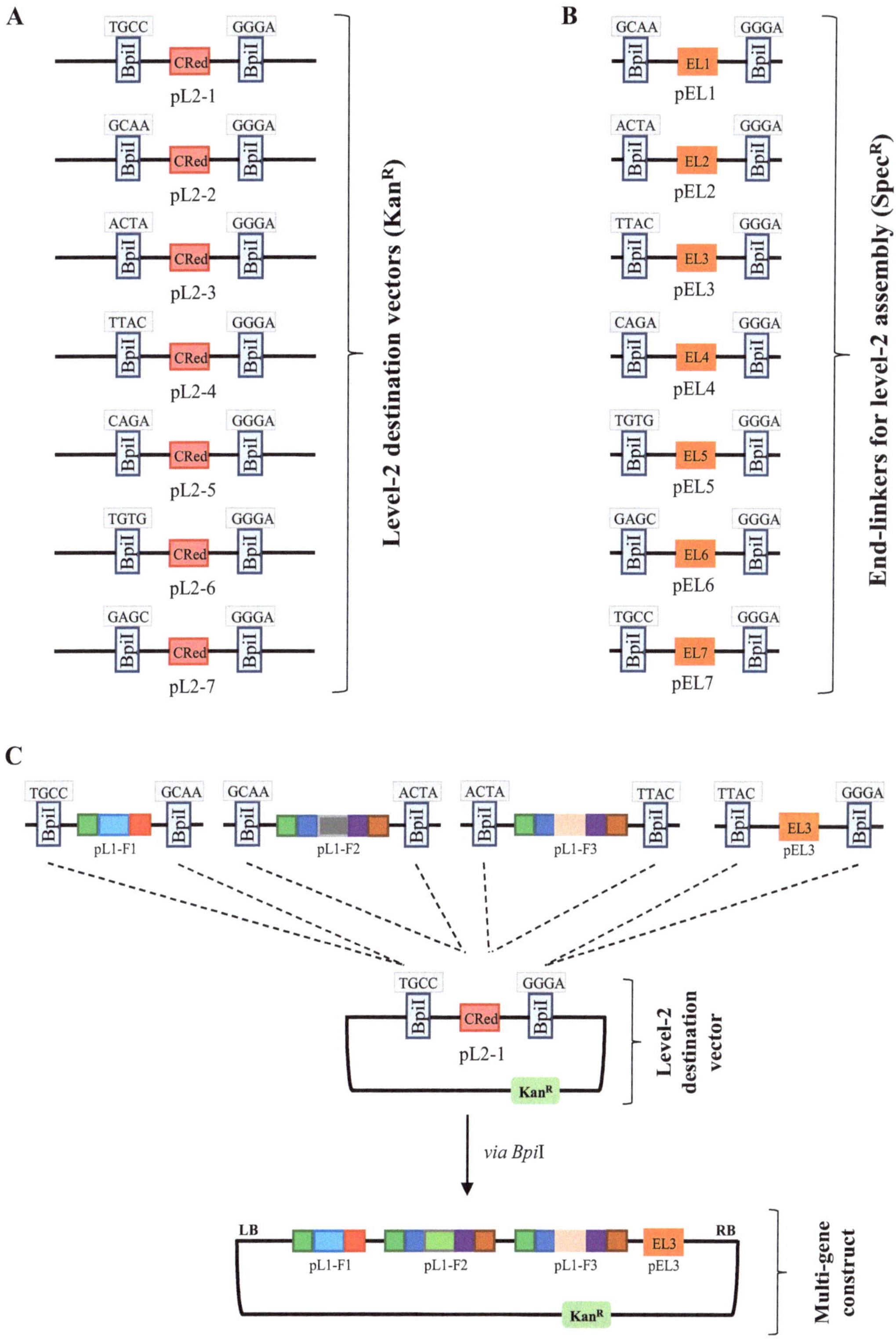

Fig. 3 Golden Gate assembly of different level-1 modules into universal level-2 destination vector. (**a**) Level-2 destination vectors. (**b**) End-linkers used to link level-1 modules to level-2 destination vector. (**c**) Cloning strategy of level-1 modules into level-2 destination vector via *Bpi*I and T4 DNA ligase following standard restriction/ligation protocol

4. Colonies exhibiting the expected white coloration were picked for plasmid isolation, purification, and subsequently screened to confirm the presence of correct constructs (*see* **Note 8**).

3.4 Golden Gate Assembly Reaction

1. All Golden Gate assembly reactions were performed in a single tube using a "one-pot" digestion-ligation strategy. A typical 15 μL reaction mixture contained:
   - 100–200 ng of acceptor plasmid
   - 20–50 ng of each DNA insert (*see* **Note 9**)
   - 1.5 μL of T4 DNA ligase buffer (10X) (*see* **Note 10**)
   - 1.5 μL of Bovine serum albumin (BSA, 10X) (*see* **Note 11**)
   - 0.5 μL of either *Bpi*I or *Bsa*I, 10 U/μL (*see* **Note 12**)
   - 0.5 μL of T4 DNA ligase, 400 U/μL (*see* **Note 10**) (*see* **Note 12**)
   - Nuclease-free water to a final volume of 15 μL (*see* **Note 13**).
2. The reaction was set up in a thermocycler as follows:

   37 °C for 20 s

   $$\left.\begin{array}{l}37\ ^\circ\mathrm{C}\ \text{for}\ 3\ \text{min}\\ 16\ ^\circ\mathrm{C}\ \text{for}\ 4\ \text{min}\end{array}\right\}\times 26$$

   50 °C for 5 min

   80 °C for 5 min

   16 °C for 5 min

3.5 Bacterial Transformation and Construct Verification

1. Chemically competent *E. coli* cells (DH5α) were freshly made and tested for their competency (*see* **Note 14**).
2. 5 μL of the Golden Gate assembly reaction was added to 50 μL of competent *E. coli* cells. The mixture was incubated on ice for 20 min, heat-shocked at 42 °C for 60 sec, and then immediately returned to ice for 2 min.
3. Cells were resuspended in 950 μL of pre-warmed LB medium and incubated at 37 °C with shaking at 200 rpm for 1 h (*see* **Note 15**).
4. 100 μL of aliquot was plated on Luria-Bertani (LB) agar plates supplemented with appropriate antibiotics such as spectinomycin (50 μg/mL), carbenicillin (50 μg/mL), kanamycin (50 μg/mL), and other supplements like X-Gal, IPTG, and incubated overnight at 37 °C (*see* **Note 16**).
5. Individual colonies were picked and grown in 5 mL LB broth with appropriate antibiotics overnight at 37 °C. Plasmid DNA was extracted and verified by PCR, and/or sanger sequencing.

3.6 Functional Validation of Multigene Construct

1. To validate the functionality of new building blocks, we assembled and tested the expression of GFP from level-2 construct.
2. The level-1 transcriptional units at the positions pL1-F1 (contains suppressor of silencing (P19)) and pL1-F2 (contains the green fluorescent protein (GFP)) (Fig. 2c), the end-linker (pEL2), and the level-2 destination vector (pL2-1) were assembled in a single cloning reaction using T4 DNA ligase and *Bpi*I (Fig. 4a).
3. The final construct was sequence verified and was transformed into competent *Agrobacterium* cells (strain GV3101) by electroporation using a Bio-Rad Gene Pulser Xcell at 2.5 kV, 25 μF, and 200 Ω (*see* **Note 17**).
4. The electroporated cells were immediately resuspended in 1 mL of LB medium and incubated at 28 °C with shaking (200 rpm) for 3 h. Cells were then plated on LB agar plates supplemented with rifampicin (30 μg/mL) and kanamycin (50 μg/mL) and incubated at 28 °C for 3 days.
5. A single *Agrobacterium* colony was cultured in 50 mL LB medium with rifampicin and kanamycin overnight at 28 °C.
6. The overnight culture was pelleted and resuspended in infiltration buffer (10 mM MES, pH 5.6; 10 mM $MgCl_2$; 200 μM acetosyringone) to a final OD_{600} of 0.2 (*see* **Note 18**).
7. The *Agrobacterium* suspension was infiltrated into the abaxial side of *Nicotiana benthamiana* leaves using a 1 mL needleless syringe, ensuring the solution spread evenly across the leaf tissue.
8. Infiltrated plants were returned to normal growth conditions. Expression of GFP (fluorescent protein) was typically observed 3–5 days post-infiltration (dpi) using UV illumination.
9. Further to compare the effectiveness of new building blocks, the coding sequence of GFP was cloned into plant binary vector pEAQ-*HT*-DEST1, serving as a positive control (Fig. 4b).

4 Notes

1. All DNA parts should be free from internal *Bsa*I and *Bpi*I recognition sites, except where necessary for DNA assembly purposes.
2. Be careful when designing custom overhangs for module or part domestication, wisely review the existing overhangs used in the cloning strategy and ensure that any new overhangs consistently feature unique bases in at least three of the four positions.

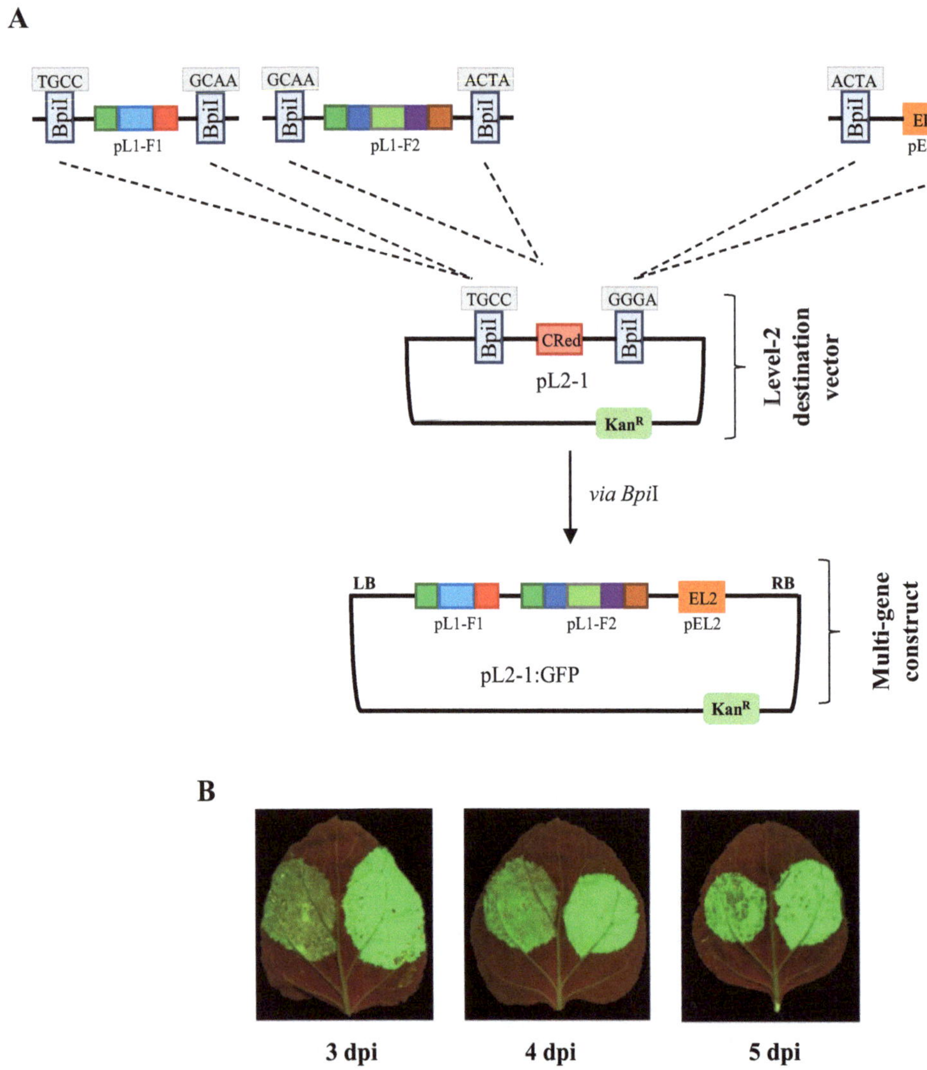

Fig. 4 Functional validation of Golden Gate level-2 multigene construct. (**a**) Strategy of assembling level-1 transcriptional units into level-2 destination vector. (**b**) Transient expression of Golden Gate level-2 multigene construct (pL2–1:GFP) and pEAQ-DEST1:GFP construct in *Nicotiana benthamiana*. The infiltrated leaves were collected at 3 dpi, 4 dpi, 5 dpi (dpi: days post infiltration) and visualized under UV illumination. The left half of the *N. benthamiana* leaf was infiltrated with pL2–1:GFP construct and the right half of the *N. benthamiana* leaf was infiltrated with pEAQ-DEST1:GFP construct

3. We have obtained good results using the Phusion High-Fidelity DNA polymerase (Thermo) with the standard 5X HF buffer.
4. Gel purification reduces template carry-through; therefore, always gel extract all PCR products and quantify the concentration using a Nanodrop spectrophotometer. Determination

of DNA concentration using the absorbance ratio of A260/280 is usually sufficient. However, if DNA assembly is inefficient, determination of DNA concentration using fluorescent dyes may be helpful in troubleshooting.

5. It should be noted that the 5′ overhang of the pL0-CDS includes a translation initiation codon (ATG). Thus, to keep coding sequence in frame, the forward primer does not need to include the native ATG. The use of multiple start codons will lead to a modified N-terminus in final expression products. Whereas, the fusion site ending with GCTT should include the native stop codon just before the 3’ GCTT overhang.
6. Validation by sanger sequencing is not necessary in most cases. However, sequencing of pL0 modules is highly recommended.
7. If your acceptor vector contained the lacZ cassette, include IPTG and X-gal for blue-white selection.
8. Plasmid purification can be done with any commercially available kit providing high-purity DNA. For L1 and L2 assemblies, prefer to do Midi-prep. Although Mini-prep DNA quality is good enough for sequencing, it is NOT generally good enough for complex DNA assemblies.
9. Always use a 2:1 molar ratio of insert:acceptor.
10. Use of the NEB ligase buffer and the NEB T4 DNA ligase works efficiently for complex assemblies.
11. If your BSA stock solution is 100X, it is always recommended to make 10X BSA solution.
12. It is highly recommended to use 0.5 μL of both restriction and ligation enzymes. Do not exceed the amount unless specifically instructed.
13. Do not exceed the final reaction volume more than 15 μL unless specifically instructed.
14. Lab-made competent cells DH5α and GV3101 are typically adequate for most purposes. However, it is recommended to test their competence in advance to ensure high cloning efficiency.
15. LB medium is used here. However, richer media or specific recovery media (e.g., SOC) may be used if transformation results in low colony numbers.
16. Antibiotics and supplements (X-Gal, IPTG) in the media may be heat sensitive. Do not add additional components until the media has cooled to at least 50 °C.
17. We tested *Agrobacterium* bacterium strain GV3101; however, other strains like LBA4404 and EHA105, might also use.
18. Do not overgrow the *Agrobacterium* culture. The optical density of culture should not exceed an OD_{600} of 1. If the final

OD_{600} is >1, the amount of dead bacterial cells will be higher which might affect the experiment. Typically, the final OD_{600} of bacterial cells in the infiltration should be in the range of 0.1 to 0.3. Additionally, always prepare fresh infiltration buffer on the day of infiltration.

Acknowledgments

A.O. would like to acknowledge the support of Biotechnological and Biological Sciences Research Council (BBSRC) Super Follow-on-Fund award BB/R005508/1, the BBRSC Institute Strategic Programme Grant 'Harnessing biosynthesis for sustainable food and health' (BB/X01097X/1), and the John Innes Foundation. We thank Professor George Lomonossoff (John Innes Centre, UK) for granting permission to use the pEAQ-*HT*-DEST1 plasmid, and JIC Horticultural Services for their assistance with plant cultivation.

References

1. Liu W, Stewart CN Jr (2015) Plant synthetic biology. Trends Plant Sci 20:309–317
2. Hughes RA, Ellington AD (2017) Synthetic DNA synthesis and assembly: putting the synthetic in synthetic biology. Cold Spring Harb Perspect Biol 9:a023812
3. Stevenson J, Krycer JR, Phan L, Brown AJ (2013) A practical comparison of ligation-independent cloning techniques. PLoS One 8:e83888
4. Xiong X, Lu Z, Ma L, Zhai C (2023) Applications of programmable endonucleases in sequence- and ligation-independent seamless DNA assembly. Biomolecules 13:1022
5. Gibson DG, Glass JI, Lartigue C, Noskov VN, Chuang RY, Algire MA et al (2010) Creation of a bacterial cell controlled by a chemically synthesized genome. Science 329:52–66
6. Oost JVD, Patinios C (2023) The genome editing revolution. Trends Biotechnol 41: 396–409
7. Chao R, Yuan Y, Zhao H (2015) Recent advances in DNA assembly technologies. FEMS Yeast Res 15:1–9
8. Engler C, Kandzia R, Marillonnet S (2008) A one pot, one step, precision cloning method with high throughput capability. PLoS One 3: e3647
9. Bird JE, Wright JM, Giachino A (2022) A user's guide to golden gate cloning methods and standards. ACS Synth Biol 11:3551–3563
10. Marillonnet S, Grützner R (2020) Synthetic DNA assembly using Golden Gate cloning and the hierarchical modular cloning pipeline. Curr Protoc Mol Biol 130:e115
11. Pingoud A, Wilson GG, Wende W (2014) Type II restriction endonucleases-a historical perspective and more. Nucleic Acids Res 42: 7489–7527
12. Sikkema AP, Tabatabaei SK, Lee YJ, Lund S, Lohman GJS (2023) High-complexity one-pot Golden Gate assembly. Curr Protoc 3:e882
13. Taylor GM, Mordaka PM, Heap JT (2019) Start-stop assembly: a functionally scarless DNA assembly system optimized for metabolic engineering. Nucleic Acids Res 47:e17
14. Werner S, Engler C, Weber E, Gruetzner R, Marillonnet S (2012) Fast track assembly of multigene constructs using Golden Gate cloning and the MoClo system. Bioengineered 3: 38–43
15. Engler C, Youles M, Gruetzner R, Ehnert TM, Werner S, Jones JD, Patron NJ, Marillonnet S (2014) A golden gate modular cloning toolbox for plants. ACS Synth Biol 3:839–843
16. Sainsbury F, Thuenemann EC, Lomonossoff GP (2009) pEAQ: versatile expression vectors for easy and quick transient expression of heterologous proteins in plants. Plant Biotechnol J 7:682–693
17. Patron NJ (2016) DNA assembly for plant biology. Curr Protoc Plant Biol 1:604–616

18. Kabadi AM, Ousterout DG, Hilton IB, Gersbach CA (2014) Multiplex CRISPR/Cas9-based genome engineering from a single lentiviral vector. Nucleic Acids Res 42:e147
19. Bortesi L, Fischer R (2015) The CRISPR/Cas9 system for plant genome editing and beyond. Biotechnol Adv 33:41–52
20. Gibson DG, Young L, Chuang RY, Venter JC, Hutchison CA III, Smith HO (2009) Enzymatic assembly of DNA molecules up to several hundred kilobases. Nat Methods 6:343–345
21. Sarrion-Perdigones A, Vazquez-Vilar M, Palací J, Castelijns B, Forment J, Ziarsolo P, Blanca J, Granell A, Orzaez D (2013) GoldenBraid 2.0: a comprehensive DNA assembly framework for plant synthetic biology. Plant Physiol 162:1618–1631
22. Weber E, Engler C, Gruetzner R, Werner S, Marillonnet S (2011) A modular cloning system for standardized assembly of multigene constructs. PLoS One 6:e16765

Chapter 9

Methods for Functional Validation of Terpenoid Metabolic Clusters in *Nicotiana benthamiana* and *Aspergillus oryzae*

Zheyong Xue, Chang Li, Xue Yin, Jiaojiao Lu, and Yanhong Guo

Abstract

Terpenoids represent one of the most structurally diverse classes of natural products, with vast potential applications in medicine, food, and agriculture. However, exploiting this potential has historically been challenging due to their low abundance in natural sources and the economic inefficiency of synthetic chemistry approaches for such complex structures. Heterologous biosynthesis offers a promising solution to these limitations. To date, multiple heterologous hosts have been developed for terpenoid production. Among these, *Nicotiana benthamiana* has emerged as a prominent platform for reconstituting plant-derived natural product pathways. Its compatibility with *Agrobacterium*-mediated transient expression, a scalable and flexible method, enables rapid integration and functional analysis of genes from diverse plant species. Conversely, *Aspergillus oryzae* has gained attention as a preferred host for fungal terpenoid pathways, owing to its highly efficient fungal expression system. Herein we present methodologies for functionally validating terpenoid biosynthetic clusters in *N. benthamiana* and *A. oryzae*.

Key words Heterologous expression, Terpenoids, *Nicotiana benthamiana*, *Aspergillus oryzae*

1 Introduction

Terpenoids represent one of the most structurally diverse classes of natural products, with significant potential applications in medicine, food, and agriculture [1, 2]. However, exploiting this potential has always been challenging due to their low abundance. In addition, the structural complexity of terpenoids often renders synthetic chemistry approaches economically unviable. Heterologous biosynthesis offers a promising strategy to overcome these limitations. To date, various heterologous hosts have been developed for terpenoid production [3, 4], among which *Nicotiana benthamiana* and *Aspergillus oryzae* have emerged as prominent platforms for studying plant- and fungal-derived natural product pathways, respectively.

N. benthamiana, a wild relative of tobacco, has been increasingly utilized over the past decade for reconstituting plant natural

Hans-Wilhelm Nützmann and Ramesha Thimmappa (eds.), *Metabolic Gene Clusters: Methods and Protocols*, Methods in Molecular Biology, vol. 3054, https://doi.org/10.1007/978-1-0716-5384-5_9,

product biosynthetic pathways [3]. Its appeal stems from two key advantages, firstly, *N. benthamiana* exhibits conserved cellular compartmentalization, cofactors, and coenzymes with other plants, allowing the direct reconstruction of foreign pathways without extensive system optimization. Secondly, the plant is highly amenable to *Agrobacterium*-mediated transient expression, a scalable method in which *Agrobacterium tumefaciens* (harboring target genes in a binary vector) is syringe-infiltrated into leaves, enabling rapid and flexible gene expression.

The filamentous fungus *A. oryzae* has been widely used in East Asian food fermentation for centuries [5]. With the development of the genetically tractable strain *A. oryzae* NSAR1, this organism has emerged as an exceptional heterologous expression platform for secondary metabolite biosynthesis [5, 6]. First of all, *A. oryzae* naturally produces low levels of endogenous secondary metabolites, minimizing interference with heterologous expressed pathways. Furthermore, the fungus exhibits robust metabolic flux capable of supplying ample precursors for various natural products, including polyketides and terpenoids, as evidenced by the high yields of heterologous expression products. Finally, unlike prokaryotic or yeast systems, *A. oryzae* can directly express intact fungal biosynthetic genes, including those containing introns, significantly simplifying genetic manipulation procedures. By using this strain, researchers can easily validate functions of fungal metabolic clusters and access their biosynthetic pathways [7]. Herein we present methodological approaches for the functional characterization of terpenoid biosynthetic gene clusters in both *N. benthamiana* and *A. oryzae*.

2 Materials

2.1 Materials for Transformation in N. benthamiana

2.1.1 Plant Expression Vector

1. pEAQ-HT vector

2.1.2 N. benthamiana Seeds

2.1.3 A. tumefaciens Transformation Reagent

1. *A. tumefaciens* strain GV3101. Other disarmed *Agrobacterium* strains, such as EHA105 and LBA4404, can also be used. However, GV3101 is the most suitable strain for *N. benthamiana* and other Solanaceae family plants.

2. Luria-Bertani (LB) medium liquid capsules: LB medium liquid capsules are added to sterile water by following the supplier recommendations and autoclaved. This can be stored at room temperature (~21 °C) for 3–4 weeks.
3. LB medium agar capsules: LB medium agar capsules are added to sterile water according to the supplier's recommendations and then autoclaved. If needed, antibiotics are added, warm medium is poured into appropriately sized Petri plates and the agar is allowed to solidify. The plates can be stored at 4 °C for 1–2 weeks.
4. Antibiotics: kanamycin (50 μg/mL), gentamycin (50 μg/mL), rifampicin (25 μg/mL) (*see* **Note 1**).

2.1.4 Agent for Infecting Plants

1. 2-(N-Morpholino) ethane sulfonic acid (MES) (1 M stock prepared in double-deionized water, adjusted to pH 5.6 with KOH).
2. 3′,5′-Dimethoxy 4′-hydroxy acetophenone (acetosyringone) (200 mM stock in dimethyl sulfoxide (DMSO) or pure ethanol).
3. Magnesium chloride ($MgCl_2$) (1 M stock prepared in double-deionized water).
4. Infiltration buffer (IB): 10 mM MES; 10 mM $MgCl_2$; 250 μM acetosyringone in double-deionized water.
5. 1 mL syringes.

2.1.5 Metabolite Identification Reagent

1. Mineral salt: NaOH, KOH.
2. Organic reagent: *n*-hexane, petroleum ether, methanol, ethyl acetate.

2.2 Materials for Transformation in A. oryzae

2.2.1 Plasmid Preparation

1. Plasmid pTAex3 or pAde A.
2. Luria-Bertani (LB) medium: Tryptone 1%; Yeast Extract 0.5%; NaCl 1%; Agar 1.5%.
3. *E. coli* DH5α cells.

2.2.2 Protoplasts Preparation

1. (Dextrin-polypeptone-yeast) DPY medium: dextrin 2%; polypeptone 1%; yeast extract 0.5%; $MgSO_4{\cdot}7H_2O$ 0.05%; KH_2PO_4 0.5%.
2. Czapek-Dox (CD)—starch medium: starch 2%; polypeptone 1%; KCl 0.2%; $NaNO_3$ 0.3%; $MgSO_4{\cdot}7H_2O$ 0.05%; KH_2PO_4 0.1%; $FeSO_4{\cdot}7H_2O$ 0.002%.
3. M medium: NH_4Cl 0.2%; $(NH_4)_2SO_4$ 0.1%; KCl 0.05%; NaCl 0.05%; KH_2PO_4 0.1%; $MgSO_4{\cdot}7H_2O$ 0.05%; $FeSO_4{\cdot}7H_2O$ 0.002%; sorbitol 21.86%; glucose 2%; agar 0.8% (upper)/1% (sublayer); adjust to pH 5.5, sterilization at 121 °C for 30 min.

4. pTAex3 selection medium: M medium + 0.15% Methionine + 0.01% Adenine.
5. pAdeA selection medium: M medium + 0.15% Methionine + 0.1% Arginine.
6. Transformation (TF) solution 0: Maleic acid 0.58% adjust to pH 5.5, sterilization at 121 °C for 30 min.
7. TF solution 1: Yatalsase 1%; $(NH_4)_2SO_4$ 7.9%; dissolved with TF solution 0, adjust the volume to the appropriate level, and sterilize through a 0.22 μm membrane filter.
8. TF solution 2: Sorbitol 21.85% (w/v); $CaCl_2{\cdot}2H_2O$ 0.735% (w/v); NaCl 0.205% (w/v); 1 M Tris-HCl (pH 7.5) 1% (v/v), sterilization at 121 °C for 30 min.

2.2.3 PEG-Mediated Protoplast Transformation

1. TF solution 3: PEG4000 60% (w/v); $CaCl_2{\cdot}2H_2O$ 0.735% (w/v); 1 M Tris-HCl (pH 7.5) 1% (v/v); sterilization at 121 °C for 30 min.

3 Methods

3.1 Methods for Transformation in N. benthamiana

3.1.1 Growing Plants

N. benthamiana plants were grown in greenhouse maintained at 26 ± 2 °C with a 12 h light/12 h darkness cycle.

3.1.2 Transforming Agrobacterium with pEAQ-Gene Constructs

1. The amplified candidate genes were successfully cloned into the pEAQ-HT-DEST1 expression vector. For reconstitution of the avenacin pathway in *N. benthamiana*, expression constructs were made using Golden Gate technology. The assembly reaction mixtures were directly used to transform *Escherichia coli* DH5α cells, which were then screened for successful clones through sequence identification. The pEAQ vectors are a series of small binary vectors designed for controlled transient expression of multiple heterologous proteins in plants.
2. Separately transform 1 μL of pEAQ-gene plasmid DNA (concentration 15 ng/μL) of each construct into *A. tumefaciens* strain GV3101 (50 μL of competent cells per plasmid) by heat shock or electroporation depending on host plant.
3. Separately set up 1 mL of culture in LB liquid medium (without antibiotics) for each transformation event. Grow the cultures at 28 °C in a shaker at 200 rpm for 2 h (*see* **Note 2**).
4. Separately spread 100 μL of each culture on LB agar plates containing rifampicin (10 μg/mL) and kanamycin (50 μg/mL). Incubate the plates at 28 °C for 2 days.

5. Set up the cultures by inoculating 1 mL of LB liquid medium containing rifampicin (10 μg/mL) and kanamycin (50 μg/mL) with a single *Agrobacterium* colony carrying the appropriate construct. Grow the cultures at 28 °C in a shaker at 200 rpm for 10–12 h.
6. Confirm the presence of pEAQ-gene carrying the target gene sequence by colony PCR analysis (*see* **Note 3**).

3.1.3 Growing Agrobacterium *Cultures and Preparation for Agro-inoculation*

1. Subculture 0.5 mL of each 1 mL culture into 10 mL of LB liquid medium containing rifampicin (10 μg/mL) and kanamycin (50 μg/mL), and grow the culture at 28 °C in a shaker at 200 rpm for 12–16 h.
2. Measure the absorbance of each culture with a spectrophotometer (OD_{600}). An OD_{600} of 0.5–0.6 is preferred.
3. When the appropriate OD is reached, collect the cells from each culture by centrifugation (5000 rpm) for 5 min at room temperature.
4. For each culture, resuspend the cells in 10 mL of *Agrobacterium* infiltration buffer, and incubate them at room temperature for 3 h.
5. Measure the absorbance with a spectrophotometer.

3.1.4 Infiltration of Agrobacterium *Into* N. benthamiana

1. Adjust the absorbance of the culture to a final OD_{600} of 0.2. For co-expression of several genes, OD_{600} of 0.2 cultures of strains expressing different genes were mixed 1:1 prior to infiltration.
2. The strains were slowly infiltrated into the abaxial side of 5–6-week-old *N. benthamiana* leaves using a 1 mL syringe until the entire leaf was infiltrated (*see* **Note 4**).
3. The plants were then maintained under greenhouse conditions. Leaves were collected approximately 6 days post-infiltration and freeze-dried using a vacuum freeze-dryer.

3.1.5 Metabolite Identification

1. Non-polar terpenoids require alkaline hydrolysis followed by extraction with *n*-hexane or petroleum ether. This approach simultaneously enables defatting by removing lipid-soluble impurities like chlorophyll through *n*-hexane extraction.
2. Polar terpenoids are isolated using aqueous methanol.
3. Among enhancement techniques, ultrasonic-assisted extraction emerges as the preferred method owing to its low-temperature operation and high efficiency. Although microwave-assisted extraction offers superior efficiency, it demands precise temperature optimization.

4. For volatile mono- and sesquiterpenes, steam distillation facilitates essential oil preparation, albeit with potential thermal degradation of sensitive components.
5. Modern supercritical CO_2 extraction excels through its low-temperature, eco-friendly operation; by modulating pressure and adding ethanol as a cosolvent, it selectively extracts terpenoids across volatile to medium-polarity ranges, establishing itself as the premium choice for advanced research.
6. Resolution of complex matrices necessitates progressive separation and purification. Initial processing employs liquid-liquid extraction for polarity-based fractionation (e.g., petroleum ether→ethyl acetate→*n*-butanol) or gel chromatography (Sephadex LH-20) to eliminate pigments and macromolecular impurities. Core separation relies on column chromatography: normal-phase silica columns with petroleum ether-ethyl acetate gradients broadly accommodate most terpenoids; reversed-phase C18 columns (water-methanol/acetonitrile systems) excel with polar terpenoids and glycosides; while alumina columns specialize in aldehyde/ketone separation.
7. Thin-layer chromatography (TLC) enables rapid crude extract screening, separation monitoring, elution optimization, and small-scale isolation (preparative TLC) for preliminary identification using visualization reagents like vanillin-sulfuric acid that yield characteristic terpenoid colors.
8. Metabolite analysis typically combines chromatographic separation with mass spectrometry. HPLC (particularly reversed-phase C_{18}) constitutes the definitive tool for compound isolation, whereas GC specializes in volatile terpenoid separation, often coupled with MS for simultaneous analysis. Novel metabolites are identified by comparing control samples (e.g., wild-type tobacco extracts) with modified samples (e.g., enzyme gene-transfected tobacco leaves). In complex mixtures, new low-abundance metabolite peaks may not be immediately evident in chromatograms, yet prior knowledge of core molecular structures and characteristic ion fragmentation patterns enables direct detection of peaks representing related compounds. Following novel peak identification, mass spectrometry techniques provide critical structural insights. Tandem mass spectrometry with exact mass measurement proves ideal for detecting modification-associated mass shifts (e.g., hydroxylation, acylation, glycosylation).
9. Final structural elucidation relies on nuclear magnetic resonance (NMR).

3.2 Methods for Transformation in A. oryzae

3.2.1 Plasmid Preparation

1. Clone each target gene in the cluster using PCR method from genomic DNA.
2. Purify the targeted genes, digest them with appropriate restriction enzymes if necessary, and ligate them into the pTAex3 vector, using an In-fusion® HD Cloning Kit according to the protocols [8]. For the introduction of the genes into pAdeA, the fragments containing the *amyB* promoter (*PamyB*) and the *amyB* terminator (*TamyB*) were amplified from the pTAex3-based plasmids, and ligated into the pAdeA vector by using an In-fusion® HD cloning kit (*see* **Note 5**).
3. Remove a vial of competent DH5α cells from −80 °C and immediately place it on ice for several minutes until just thawed. Gently mix by tapping (*see* **Note 6**).
4. Gently add the ligated mixture to the cells. Swirl the tube gently or flick it a few times. Incubate the mixture on ice for 30 min.
5. Rapidly transfer the tubes from ice to 42 °C for exactly 45 s. Immediately return the tubes to ice for 2 min.
6. Add 450 μL of sterile LB broth (without antibiotic) to each tube. Incubate in a 37 °C shaking incubator for 60 min.
7. Spread 200 μL of the transformation culture onto LB agar plates containing the appropriate antibiotic.
8. Let the plates dry, then incubate them upside down at 37 °C for 12–16 h (overnight), extract the plasmids next day (*see* **Note 7**).

3.2.2 Protoplasts Preparation

1. Inoculate 10 mL of DPY broth with ~10^6 spores of *A. oryzae* NSARl [9, 10] at 30 °C for 1–2 days, then incubate 100 mL of DPY broth with the pre-cultured *A. oryzae* at 30 °C with vigorous shaking (120–150 rpm) for 24 h (*see* **Note 8**).
2. Prepare 10 mL of TF Solution 1, and filter it using a 0.22 μm filter membrane into a 50 mL centrifuge tube.
3. Filter 30 mL of the *A. oryzae* culture through Bio-sterilized cotton mesh to collect the mycelium. Gently blot dry with sterile paper towels. Put them in a centrifuge tube with TF solution 1. Add pre-warmed (30 °C) enzyme solution. Incubate at 30 °C with gentle shaking (50–80 rpm) for 2–3 h (*see* **Note 9**).
4. Filter the digest through Bio-sterilized cotton into a 50 mL tube to remove undigested hyphae. Add 10 mL of TF solution 2, gently centrifuge the filtrate at 1500–2000 g for 10 min at 4 °C to pellet protoplasts. Carefully discard the supernatant. Gently resuspend the pellet in TF solution 2 buffer. Wash by gentle centrifugation and finally resuspend in 5 mL of TF solution 2 (*see* **Note 10**).

5. Count protoplasts with a hemocytometer (aim for 10^7–10^8 protoplasts/mL).

3.2.3 PEG-Mediated Protoplast Transformation of A. oryzae

1. Transfer 200 μL of the protoplast solution to a 15 mL centrifuge tube using a 1 mL pipette tip. Add 10 μL of plasmid at a concentration of 10 μg/10 μL and mix gently. Incubate the mixture on ice for 30 min. Then, add TF solution 3 to the suspension in three aliquots of 250 μL, 250 μL, and 850 μL, respectively. After each addition, gently pipette up and down using a 1 mL pipette tip. Finally, let the mixture stand at room temperature for 20 min (*see* **Note 11**).
2. Add 5 mL of TF solution 2, invert the tube several times, and mix gently. Centrifuge at 1500 rpm for 10 min, then discard the supernatant. Add 200 μL of TF solution 2 and gently resuspend the pellet using a 1 mL pipette tip, pour them into the sublayer medium. Subsequently, quickly add 5 mL of pre-warmed (50 °C) upper medium and mix rapidly (*see* **Note 12**).
3. After the plate surface is fully dried, seal it with parafilm and incubate for 3–7 days. Subculture the obtained transformants on selection medium without sorbitol for 1–3 rounds to allow the transformants to stabilize (*see* **Note 13**).
4. Select 3–5 strains and culture them in 100 mL CD-starch (30 °C, 150 rpm) for 6 days. Collect the mycelium by filtration through a funnel, then fully absorb the surface moisture of the mycelium with absorbent paper. Subsequently, soak the mycelium in acetone overnight, and repeat this operation twice. Combine the three acetone extracts, concentrate under reduced pressure, dissolve in methanol, and prepare for HPLC analysis. Extract the culture medium once with an equal volume of ethyl acetate, concentrate under reduced pressure, dissolve in methanol, and prepare for HPLC analysis.
5. Analyze the products from each of the transformants, for example, with a solvent system of 0.5% acetic acid (solvent A) and acetonitrile containing 0.5% acetic acid (solvent B), at a flow rate of 1.0 mL/min and a column temperature of 40 °C (Fig. 1). Separation was performed with solvent B/solvent A (10:90) for 5 min, a linear gradient from 10:90 to 100:0 within the following 20 min, followed by 100:0 for 5 additional min.

4 Notes

1. Kanamycin and gentamycin are all dissolved in double-deionized water, while rifampicin is dissolved in DMSO. All antibiotics should be filtered before use.

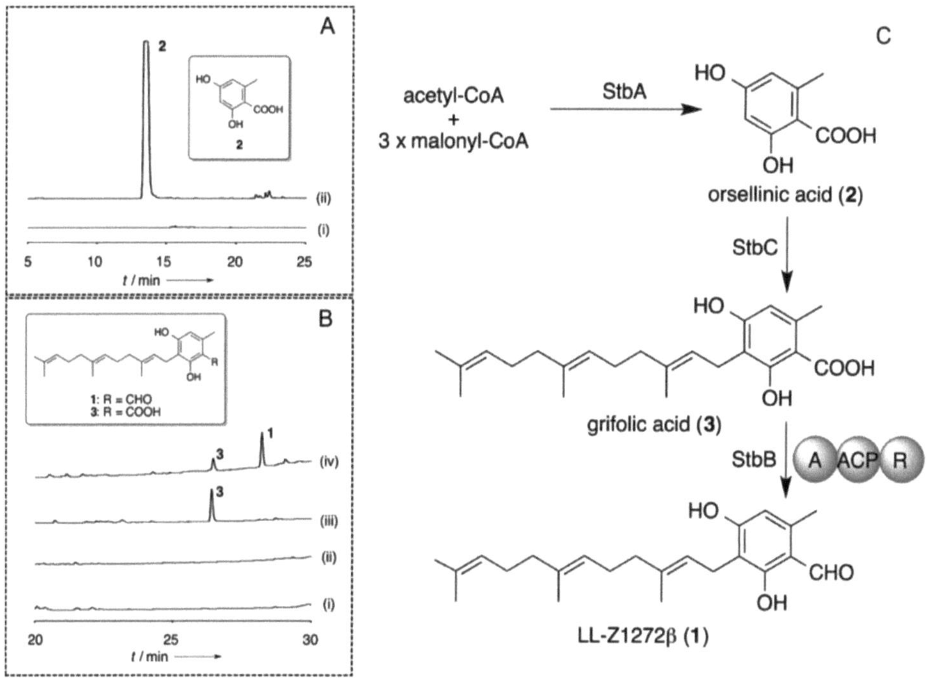

Fig. 1 (**a**) HPLC profile of the culture supernatant extract of *A. oryzae* transformants harboring (i) empty vector and (ii) *stbA* (254 nm). (**b**) HPLC profile of the mycelial extract of *A. oryzae* transformants harboring (i) *stbA*, (ii) *stbA* and *stbB*, (iii) *stbA* and *stbC*, and (iv) *stbA*, *stbB*, and *stbC* (254 nm). (**c**) Proposed biosynthetic pathway of LL-Z1272β

2. Growing temperatures for *Agrobacterium* should not exceed 28 °C as loss of the helper plasmid may occur. The volume of culture to be grown depends on the number of plants to be infiltrated.
3. Individual bacterial colonies carrying each construct can be streaked on an LB agar plate containing antibiotics and grown as described in this step; after bacterial growth, these plates can be stored for 1 week at 4 °C. Liquid cultures of bacteria can be stored as glycerol stocks for several years.
4. Cultures can accidentally produce spray during infiltration. Use a face shield or other appropriate face protection, and wear a lab coat.
5. For In-fusion, especially the molar ratio (insert and vector), please follow the kit instructions precisely.
6. Thaw slowly on ice. Never warm by hand or at room temperature. Handle gently to avoid vortexing or vigorous pipetting. Please use immediately, because once thawed, cells lose competence rapidly.
7. Always plate a sample from the "no-DNA" control transformation to confirm there is no contamination.

8. Use a hemocytometer to accurately count spores to ~1×10^6. Too many spores lead to overly dense growth, too few extend lag phase.
9. Gentle shaking (50–80 rpm) is critical for mixing without shearing fragile protoplasts. Monitor digestion visually (microscope) after 2 h, which should appear as small, spherical, refractive cells. Over-digestion (3–4 h) reduces viability.
10. Resuspend in 5 mL TF Solution 2, it could determine the protoplast concentration.
11. Gently by flicking or slow pipetting, avoid bubbles or shearing. Adding in three gradual aliquots of TF Solution 3 is critical to reduce osmotic shock.
12. The upper medium must be pre-warmed to 50 °C to keep it liquid but not so hot as to kill protoplasts.
13. Plate onto fresh selection plates without sorbitol to encourage cell wall regeneration and stabilize integration.

References

1. Luo P, Huang JH, Lv JM et al (2024) Biosynthesis of fungal terpenoids. Nat Prod Rep 41:748–783
2. Christianson DW (2017) Structural and chemical biology of terpenoid cyclases. Chem Rev 117:11570–11648
3. Reed J, Osbourn A (2018) Engineering terpenoid production through transient expression in *Nicotiana benthamiana*. Plant Cell Rep 37:1431–1441
4. Yang H, Song CN, Liu CW et al (2024) Synthetic biology tools for engineering *Aspergillus oryzae*. J Fungi 10:34
5. Dong JY, Li M, Xiao ZH et al (2022) Recent advances in heterologous production of natural products using *Aspergillus oryzae*. Synth Biol J 3(6):1126–1149
6. Awakawa T, Abe I (2021) Reconstitution of polyketide-derived meroterpenoid biosynthetic pathway in *Aspergillus oryzae*. J Fungi 7:486
7. Li XY, Lv JM, Cao ZQ et al (2023) Biosynthetic characterization of the antifungal fernane-type triterpenoid polytolypin for generation of new analogues via combinatorial biosynthesis. Org Biomol Chem 21(4):851–857
8. Cao ZQ, Li SY, Lv JM et al (2019) Biosynthesis of clinically used antibiotic fusidic acid and identification of two short-chain dehydrogenase/reductases with converse stereoselectivity. Acta Pharm Sin B 9(2):433–442
9. Li C, Matsuda Y, Gao H et al (2016) Biosynthesis of LL-Z1272β: discovery of a new member of NRPS-like enzymes for aryl-aldehyde formation. Chembiochem 17(10):904–907
10. Lv JM, Hu D, Gao H et al (2018) Biosynthesis of helvolic acid and identification of an unusual C-4-demethylation process distinct from sterol biosynthesis. Nat Commun 8(1):1644

Chapter 10

Characterization of Bioactive Saponins from Sea Cucumbers

Karuna Bhaskaran, Rajesh Chandra Misra, Anne Osbourn, and Ramesha Thimmappa

Abstract

Sea cucumbers are highly valued for their bioactive constituents, particularly triterpenoid saponins, which exhibit diverse pharmacological properties. The complex amphiphilic nature of saponins poses significant challenges in their extraction and purification. Their structural diversity and amphiphilicity demand carefully optimized protocols to maximize yield and purity. Moreover, the variability of saponin profiles between different sea cucumber species and tissues necessitates reproducible methods that can be scaled from small laboratory samples to larger quantities suitable for industrial applications. This chapter describes comprehensive, reproducible protocols for both small- and large-scale extraction of saponins from sea cucumber tissues, followed by chromatographic separation and bioactivity-guided fractionation. Subsequently, analytical techniques such as thin-layer chromatography (TLC), yeast growth inhibition assays, and liquid chromatography–mass spectrometry (LC–MS) are detailed. These detailed extraction, purification, and analytical protocols facilitate downstream applications such as structural characterization, quantification, and functional studies of saponins.

Key words Sea cucumber, Triterpenoid saponins, Extraction, Flash chromatography, TLC, LC–MS

1 Introduction

Sea cucumbers (holothurians) have been an integral part of traditional medicine and dietary practices for centuries, especially within Asian countries such as China, Japan, Korea, and Southeast Asian nations. Out of approximately 1500 known sea cucumber species worldwide, around 100 are utilized directly for human consumption, with more than 40 species recognized as edible and commercially exploited. Historically, they have been associated with wound healing, boosting immunity, and alleviating various ailments—reflecting their significant role in folk medicine systems [1, 2]. In recent decades, scientific research has intensified around sea

Hans-Wilhelm Nützmann and Ramesha Thimmappa (eds.), *Metabolic Gene Clusters: Methods and Protocols*, Methods in Molecular Biology, vol. 3054, https://doi.org/10.1007/978-1-0716-5384-5_10,

cucumbers due to their rich repository of bioactive compounds, such as chondroitin sulfates, sulfated polysaccharides, omega fatty acids, and triterpenoid glycosides (saponins) [3, 4]. These compounds are notable for their potent biological activities, including anticancer, antiviral, antifungal, anti-inflammatory, and antioxidant activities [5, 6]. Beyond their medicinal potential, sea cucumbers are highly nutritious. They contain high-quality proteins, beneficial lipids, vitamins, a variety of trace elements, and minerals such as iron, calcium, zinc, and magnesium, essential for human health [2]. Among the bioactive constituents, saponins stand out as major contributors to the pharmacological effects attributed to sea cucumbers. Sea cucumbers (Class Holothuroidea) are unique among marine invertebrates in producing structurally diverse triterpenoid glycosides that function as chemical defense compounds [3]. These saponins display a broad range of biological activities, including antifungal, cytotoxic, and immunomodulatory effects [4, 5]. Despite their potential, isolating and purifying saponins from sea cucumbers pose significant technical challenges. Due to their amphiphilic nature, they tend to co-extract with lipids, sterols, and other hydrophobic molecules present in the tissue, complicating purification processes. Techniques such as solvent extraction, chromatography, and advanced purification methods are often required, but these processes can be laborious and resource-intensive [7]. This chapter describes standardized, step-by-step protocols for the extraction, purification, bioactivity assessment, and analytical characterization of sea cucumber saponins. The methods are optimized for reproducibility and are suitable for application in molecular and chemical biology laboratories, in accordance with the Methods in Molecular Biology format.

2 Materials

2.1 Biological Materials

1. Sea cucumber (*Parastichopus parvimensis*; WK Distribution) or freshly collected specimens.
2. *Saccharomyces cerevisiae* strain Y21900.
3. Artificial seawater aquarium (salinity 28%; 25 °C; pH 8.0; 12 h light/12 h dark cycle).

2.2 Reagents

1. Ethanol (70–100%, molecular biology grade).
2. Methanol (HPLC grade).
3. Chloroform.
4. *n*-Hexane.
5. *n*-Butanol.
6. Ethyl acetate.

7. Acetonitrile.
8. Dichloromethane (DCM).
9. Ultra-pure water (18.2 MΩ; Milli-Q).
10. Formic acid (LC–MS grade).
11. TLC visualization reagent: acetic acid: sulfuric acid: *p*-anisaldehyde (48:1:1, v/v/v).

2.3 Consumables

1. Whatman No. 1 filter paper (11 μm).
2. TLC plates (silica gel 60 F254, glass or aluminum-backed).
3. Silica column (Biotage SNAP Ultra 100 g, 39 × 157 mm, 25 μm).
4. LC–MS vial inserts.
5. 96-well microplates.
6. Yeast extract–peptone–dextrose (YPD) medium: 1% (w/v) Yeast extract, 2% (w/v) peptone, 2% (w/v) dextrose.
7. Glass bottles.
8. Glass test tubes.

2.4 Equipment

1. Freeze dryer (VirTis BenchTop K).
2. Rotavapor R-300 (BUCHI).
3. Centrifuge.
4. Incubator shaker.
5. Automated flash chromatography system (Isolera Biotage).
6. TLC plate heater (CAMAG).
7. FLUOstar Omega multidetector microplate reader (BMG LABTECH).
8. Q Exactive Plus Hybrid Quadrupole–Orbitrap mass spectrometer (Thermo Scientific) coupled to a UHPLC system.

3 Methods

3.1 Extraction and Liquid–Liquid Partitioning of Saponins from P. parvimensis

1. *Sample preparation*: Processed sea cucumbers (*P. parvimensis*, ~1.75 kg dry weight; WK Distribution) were soaked in distilled water for 48 h at room temperature to facilitate tissue swelling and softening (Fig. 1) (*see* **Note 1**).
2. *Tissue pulverization*: The softened tissues were thoroughly ground to produce a homogeneous mass.
3. *Ethanolic extraction*: The pulverized tissue was soaked in 100% ethanol (v/v) for 2 weeks at room temperature to extract saponins and related metabolites (*see* **Note 2**).

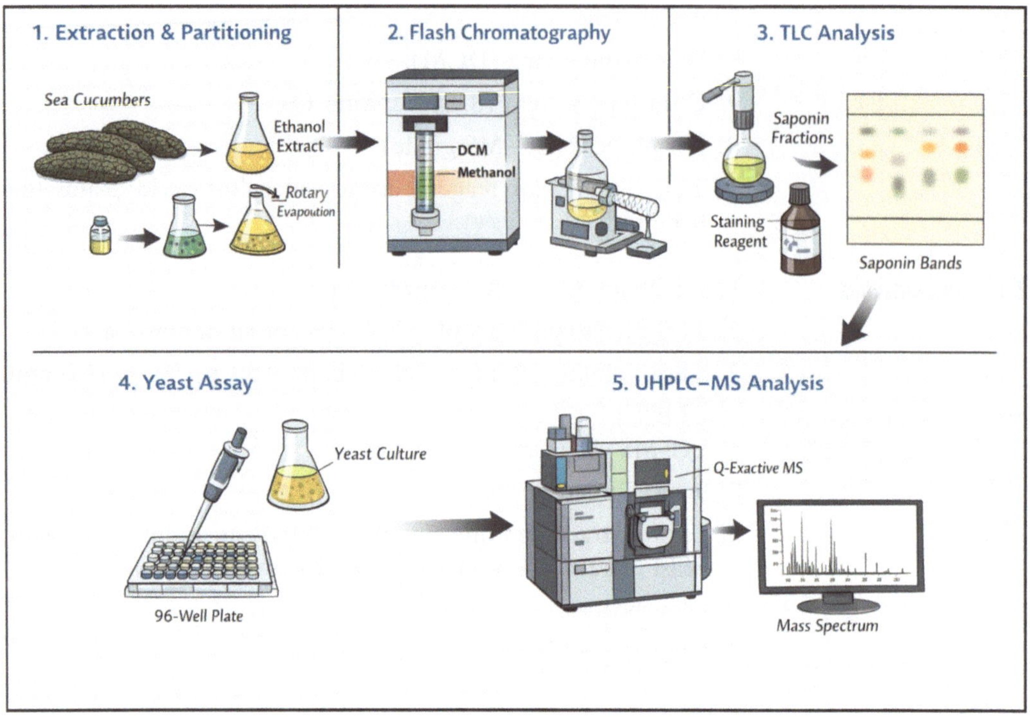

Fig. 1 Workflow for extraction, purification, and characterization of sea cucumber saponins. A schematic overview of the experimental pipeline used for the isolation and analysis of triterpenoid saponins from *Parastichopus parvimensis*. (1) Processed sea cucumber tissues were subjected to ethanol extraction followed by sequential liquid–liquid partitioning using n-hexane and n-butanol to enrich saponins. (2) Crude saponin extracts were purified by automated flash chromatography using a normal-phase silica column and stepwise dichloromethane–methanol gradients. (3) Flash chromatography fractions were screened for saponins by thin-layer chromatography (TLC) using anisaldehyde-based visualization. (4) Bioactive fractions were evaluated using a yeast growth inhibition assay in 96-well plates to assess antifungal activity. (5) Saponin-containing fractions were further characterized by UHPLC–MS using high-resolution Orbitrap mass spectrometry for molecular mass determination and structural annotation. (The workflow diagram was prepared using Biorender [8])

4. *Filtration and concentration*: The ethanol extract was filtered through Whatman No. 1 filter paper and concentrated to dryness under reduced pressure using a Rotavapor R-300 (BUCHI) at 30 °C (*see* **Note 3**).
5. *n-Hexane partitioning*: The residual aqueous phase was extracted with *n*-hexane to remove non-polar compounds like sterols (*see* **Note 4**).
6. *n-Butanol partitioning*: The remaining aqueous phase was partitioned with *n*-butanol at a ratio of 1:5 (v/v) and centrifuged at 4000 rpm to facilitate phase separation (Fig. 1) (*see* **Note 5**).
7. *Repeated extraction*: *n*-Butanol extraction was repeated three times to ensure efficient recovery of saponins, and the combined *n*-butanol phases were collected (*see* **Note 6**).

8. *Drying of extracts*: Combined *n*-butanol extracts were dried under vacuum at 56 °C using a Rotavapor R-300 (BUCHI).
9. *Freeze-drying*: The resulting greasy, yellowish extract was freeze-dried to remove residual *n*-butanol and stored for further purification and analysis (*see* **Note** 7).

3.2 Flash Chromatographic Separation of Saponins

1. *Sample preparation*: approximately 3 grams of the freeze-dried extract were prepared for chromatographic separation by dry loading onto a normal-phase silica column (Biotage SNAP Ultra 100 g, 39 × 157 mm, 25 μm) (Fig. 1) (*see* **Note 8**).
2. *Chromatography system setup*: The flash chromatography was performed using an automated flash chromatographic purification system (Isolera One, Biotage).
3. *Mobile phase preparation*: Solvent A was prepared using dichloromethane (DCM), while solvent B consisted of methanol containing 7.5% (v/v) water (*see* **Note 9**).
4. *Gradient elution*: Chromatographic separation commenced with 100% solvent A (DCM), followed by a linear gradient to 15% solvent B (methanol with water) over 50 column volumes. Subsequently, the solvent B proportion was increased to 30% to elute saponin-enriched fractions (*see* **Note 10**).
5. *Fraction collection*: Fractions were collected automatically according to the system's settings and then processed further for downstream analysis.

3.3 Thin-Layer Chromatographic Analysis of Flash Chromatography Fractions

1. *Fraction collection*: During flash chromatographic separation, various fractions (22 mL) were collected, dried, and subsequently prepared for thin-layer chromatography (TLC) analysis.
2. *TLC plate preparation*: TLC analysis was conducted using silica gel 60G glass plates (20 × 20 cm, Merck).
3. *Sample application and standard*: The dried fractions were dissolved in an appropriate solvent and spotted onto TLC plates alongside avenacin A-1 from oats, which was used as a tracking standard (*see* **Note 11**).
4. *Plate development*: The plates were developed in a solvent system composed of chloroform, methanol, and water in a ratio of 65:35:10 (v/v/v), until the solvent front reached the desired height (*see* **Note 12**).
5. *Visualization of saponins*: After development, the plates were sprayed with a visualization reagent containing acetic acid, H_2SO_4, and *p*-anisaldehyde (48:1:1, v/v/v), and then heated at 120 °C for 5 min using a TLC plate heater (CAMAG) to visualize saponin bands (Fig. 1) (*see* **Note 13**).

3.4 Yeast Growth Inhibition Assay

1. *Yeast strain and culture conditions*: Yeast growth inhibition assays were conducted using *Saccharomyces cerevisiae* strain Y21900 (BY4743 background; MATa/MATα; ura3Δ0/ura3Δ0; leu2Δ0/leu2Δ0; his3Δ1/his3Δ1; met15Δ0/MET15; LYS2/lys2Δ0; YHR072w/YHR072w::kanMX4; Euroscarf). Cultures were maintained in YPD medium (*see* **Note 14**).
2. *Preparation of saponin solutions*: Stock solutions of crude extracts and saponin standards were prepared at a concentration of 1 mg mL^{-1} in methanol. Working solutions were obtained through serial dilutions in methanol to achieve desired concentrations.
3. *Sample origin and replication*: Fractions containing saponin-like compounds, as determined by thin-layer chromatography and migration proximal to the avenacin A1 reference standard, were pooled, concentrated, and used for subsequent bioactivity assays. Two technical replicates were included for each condition to account for variability associated with low-volume pipetting (*see* **Note 15**).
4. *Plate setup*: Assays were performed in 96-well skirted, round-bottom microplates. Each well received 100 μL of YPD medium, and was inoculated with 1 μL of yeast culture diluted 100-fold from an overnight culture (*see* **Note 16**).
5. *Addition of test compounds*: Saponin standards were added to achieve final concentrations of 1–100 μg mL^{-1} (1–5 μL per well), while crude extracts were added to final concentrations of 25–100 μg mL^{-1} (1–5 μL per well) at the start of the incubation (*see* **Note 17**).
6. *Incubation*: Plates were incubated overnight at 30 °C in a shaking incubator set to 100 rpm.
7. *Growth measurement*: Yeast growth was monitored by measuring optical density at 600 nm (OD_{600}) using a FLUOstar Omega multidetector microplate reader (BMG LABTECH) with default settings (*see* **Note 18**).
8. *Data normalization*: OD_{600} values were corrected by subtracting reading from wells containing only YPD medium to account for background absorbance (Figs. 1 and 2).

3.5 UHPLC–MS Analysis and Characterization of Saponins

1. *Instrument setup*: Saponin analysis was performed using a Q Exactive Plus Hybrid Quadrupole–Orbitrap mass spectrometer (Thermo Scientific) integrated with an ultra-high-performance liquid chromatography (UHPLC) system (Figs. 1 and 2).
2. *Column configuration*: Chromatography was performed using a Kinetex XB-C18 column (2.6 μm particle size, 100 Å pore size, 50 × 2.1 mm; Phenomenex) by maintaining the column temperature at 30 °C (*see* **Note 19**).

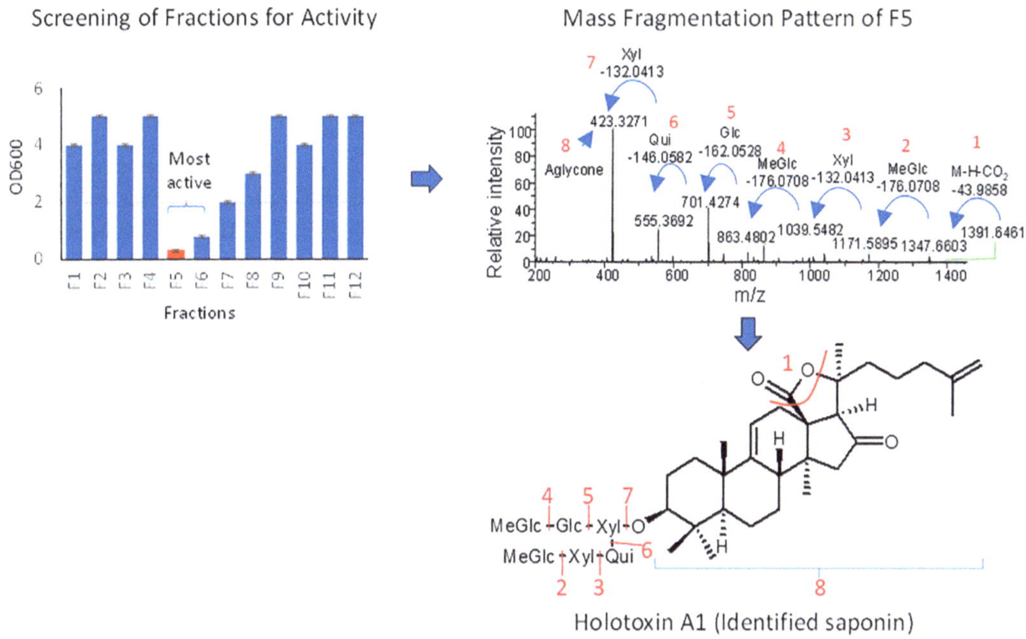

Fig. 2 Bioactivity-guided fractionation and UHPLC–MS–based identification of holotoxin A1. Flash chromatography fractions (F1–F12) obtained from *Parastichopus parvimensis* extracts were screened for antifungal activity using a yeast growth-inhibition assay (left panel). Yeast growth was quantified by measuring optical density at 600 nm (OD_{600}). Fractions F5 and F6 exhibited the strongest inhibitory activity, as indicated by markedly reduced OD_{600} values, and were therefore selected for further chemical characterization. UHPLC–MS/MS analysis of these bioactive fractions is shown in the right panels. Tandem mass spectrometry of the holotoxin A1 parent ion produced a series of diagnostic fragment ions arising from neutral losses and glycosidic bond cleavages within the sugar chain and aglycone. Numbered fragment ions correspond to m/z 43.9858 (fragment 1; $[M\text{–}H\text{–}CO_2]^-$), m/z 176.0708 (fragment 2; 3-*O*-methyl-*D*-glucose), m/z 132.0413 (fragment 3; xylose), m/z 176.0680 (fragment 4; 3-*O*-methyl-*D*-glucose), m/z 162.0528 (fragment 5; glucose), m/z 146.0582 (fragment 6; quinovose), m/z 132.0421 (fragment 7; xylose), and m/z 423.3271 (fragment 8; triterpenoid aglycone). The observed fragmentation pattern, including characteristic losses of methylated and non-methylated hexoses, deoxyhexose residues, and release of the aglycone, is consistent with previously reported MS/MS data for holotoxin A1 and supports its confident identification as the major bioactive saponin present in fraction F5

3. *Mobile phase preparation*: Prepare solvent A as water containing 0.1% (v/v) formic acid, and solvent B as acetonitrile containing 0.1% (v/v) formic acid.
4. *Chromatographic conditions*: The flow rate was set to 0.3 mL min^{-1} and an injection volume of 10 μL (*see* **Note 20**).
5. *Gradient elution program*: The following gradient program was implemented:
   (i) Linear increase from 0% to 30% solvent B for 0–5 min.
   (ii) Linear increase from 30% to 50% solvent B for 5–28 min.
   (iii) Linear increase from 50% to 100% solvent B for 28–33 min.

(iv) Linear decrease from 100% to 20% solvent B for 33–34 min.

(v) Finally, hold at 20% solvent B for additional 1 min to re-equilibrate column, giving a total run time of 35 min.

6. *Ionization and scan settings*: The mass spectra was acquired in negative electrospray ionization (ESI) mode over an *m/z* range of 100–2000 with a mass resolution of 70,000 (*see* **Notes 21** and **22**).
7. *Data-dependent acquisition*: Full MS scans in combination with data-dependent MS/MS acquisition (full MS/dd-MS^2, top three) were performed (Figs. 1 and 2).
8. *Targeted acquisition*: Selected ion monitoring (SIM) and parallel reaction monitoring (PRM) were conducted with a mass resolution of 35,000 full width at half maximum (FWHM), a mass isolation window of 1.6 *m/z*, and an automatic gain control (AGC) target of 3×10^6 (*see* **Notes 22** and **23**).
9. *Inclusion list*: In PRM mode, a predefined inclusion list containing the accurate masses of known saponins for targeted data acquisition was utilized (*see* **Note 24**).
10. *Data processing*: The data was acquired and processed using Thermo Scientific Xcalibur software (version 4.3.73.11). Compare fragmentation spectra with previously reported saponins from *P. parvimensis* and *A. japonicus* for compound identification and annotation (Figs. 1 and 2) (*see* **Note 25**).

4 Notes

1. Prolonged soaking in distilled water enhances solvent penetration and increases extraction efficiency for processed sea cucumbers.
2. An optimal solvent volume of approximately 8 mL per gram of dry tissue is recommended. Insufficient solvent may lead to incomplete extraction and lower yields.
3. Complete removal of ethanol before liquid–liquid partitioning is essential to ensure proper phase separation and optimal extraction efficiency.
4. For higher purity, an additional ethyl acetate partitioning step can be performed after *n*-hexane extraction to separate moderately polar saponins from highly polar impurities.
5. Saponins are amphiphilic that preferentially partition into *n*-butanol during liquid–liquid extraction.
6. Repeated *n*-butanol extractions significantly improve saponin recovery owing to their amphiphilic nature.

7. Freeze-dried saponin fractions should be stored at −20 °C or −80 °C in airtight containers, protected from light and moisture.
8. Prior to sample loading, silica columns should be thoroughly equilibrated with the starting mobile phase to ensure reproducible separation and to prevent band broadening during flash chromatography.
9. Dichloromethane (DCM) elutes non-polar compounds first, while increasing methanol concentration enhances elution of polar saponin; stepwise gradients improve resolution and minimize co-elution.
10. Column volumes calculations should be based on the manufacturer-specified volume for the SNAP Ultra 100 g column.
11. Capillary spotting prevents sample overloading and enhances band resolution; Avenacin A-1 serves as a reliable reference for monitoring saponin migration.
12. TLC chambers should be tightly sealed and pre-saturated with solvent vapor for at least 20 min before development to improve reproducibility; solvent systems may be optimized according to saponin polarity.
13. Overheating can cause excessive background staining; therefore, heating durations should be carefully controlled.
14. The Y21900 strain is suitable for saponin bioactivity screening due to its well-characterized genetic background and consistent growth behavior.
15. For optimal saponin recovery, freshly dissected tissues should be flash-frozen in liquid nitrogen immediately after dissection. If immediate processing is not possible, samples can be freeze-dried and stored at −80 °C for several months without significant saponin loss.
16. Use yeast cultures in logarithmic growth phase to ensure consistent growth kinetics across experiments. Including technical replicates is crucial when working with small volume to reduce variability caused by pipetting errors.
17. Control experiments with methanol-only solutions at the highest solvent concentration used should be included to rule out solvent effect on growth.
18. Growth inhibition may vary depending on saponin structure and concentration; preliminary dose-finding experiments are recommended.
19. The short column length (50 mm) allows rapid analysis but may reduce resolution for closely eluting saponins; longer columns (e.g., 100 mm) can be employed for higher separation efficiency.

20. Highly glycosylated saponins may adhere to tubing and column surfaces; flushing the system with high organic solvent (≥95% acetonitrile) between runs can minimize carryover.
21. Negative ESI mode offers superior sensitivity for triterpenoid saponins due to efficient deprotonation of glycosidic moieties; positive mode can be used as a complementary approach but generally provides lower signals.
22. Higher mass resolution improves mass accuracy but may reduce scan speed; the chosen resolutions (70,000 for full MS and 35,000 for PRM) balance sensitivity with duty cycle.
23. Collision energies for PRM should be empirically optimized for each saponin to achieve informative fragmentation while minimizing precursor ion depletion.
24. Verify the accuracy of *m/z* values in the inclusion list before data acquisition; incorrect *m/z* values can lead to missed PRM events or poor fragmentation (Fig. 2).
25. Interpret fragmentation patterns alongside retention time and accurate masses, as structural isomers may produce similar MS/MS spectra (Fig. 2).

Acknowledgments

This work was supported by DBT-Ramalingaswami grant, Govt of India to R.T. and A.O. would like to acknowledge the support of Wellcome Discovery Award #227375/Z/23/Z, the BBRSC Institute Strategic Programme Grant "Harnessing biosynthesis for sustainable food and health" (BB/X01097X/1), and the John Innes Foundation.

References

1. Matsuno T, Yamanouchi T (1961) A new triterpenoid sapogenin of animal origin (sea cucumber). Nature 191:75–76
2. Fagbohun OF, Joseph JS, Oriyomi OV, Rupasinghe HPV (2023) Saponins of North Atlantic sea cucumber: chemistry, health benefits, and future prospectives. Mar Drugs 21:262
3. Thimmappa R, Wang S, Zheng M, Misra RC, Huang AC, Saalbach G, Chang Y, Zhou Z, Hinman V, Bao Z, Osbourn A (2022) Biosynthesis of saponin defensive compounds in sea cucumbers. Nat Chem Biol 18:774–781
4. Liu Y, Lu Z, Yan Z, Lin A, Han S, Li Y, Yang X, Li X, Yin X, Zhang R, Li K (2024) Sea cucumber viscera contains novel non-holostane-type glycoside toxins that possess a putative chemical defense function. J Chem Ecol 50:185–196
5. Bahrami Y, Zhang W, Franco C (2014) Discovery of novel saponins from the viscera of the sea cucumber *Holothuria lessoni*. Mar Drugs 12: 2633–2667
6. Seo HD, Lee JY, Park SH, Lee E, Hahm JH, Ahn J, Jang AR, An SH, Ha JH, No KT, Jung CH (2024) Identification of novel anti-obesity saponins from the ovary of sea cucumber (*Stichopus japonicus*). Heliyon 10:e36943
7. Majinda RR (2012) Extraction and isolation of saponins. Methods Mol Biol 864:415–426
8. BioRender.com (2025) BioRender: scientific illustration software. https://www.biorender.com

Chapter 11

Rhizobium rhizogenes-Mediated Hairy Root Transformation Protocol for *Lotus japonicus* and Other Legumes

Panagiotis Vletsos, Jenny Koukara, and Kalliope K. Papadopoulou

Abstract

Rhizobium rhizogenes-mediated hairy root transformation has been extensively applied across various legumes, ranging from model species such as *Medicago truncatula* and *Lotus japonicus* to crop legumes like *Glycine max* and *Phaseolus vulgaris*. This technique is highly regarded for its simplicity and rapid production of transgenic roots, making it invaluable for studies on gene expression and function, root biology, symbiotic interactions, and metabolic engineering. Consequently, multiple protocols have been developed to optimize its use in legumes. In this chapter, we present a detailed protocol for hairy root transformation in the model plant *Lotus japonicus*, emphasizing key elements essential for success that can be adapted for use in other legumes.

Key words *Rhizobium rhizogenes*, Hairy root transformation, *Lotus japonicus*, Legumes

1 Introduction

Agrobacterium rhizogenes, now *Rhizobium rhizogenes*, is a Gram-negative bacterium that infects dicotyledonous plants. When wounded, plants release signals into the environment, attracting the bacterium into the wounded site and activating the transfer of the T-DNA from the Ri plasmid to the plant genome [1, 2]. T-DNA integration induces the formation of multi-branched adventitious roots at wounded sites, resulting in the characteristic hairy root phenotype [3, 4]. The *R. rhizogenes*-based transformation system is recognized for its speed, simplicity, and high efficiency [5]. The system typically employs the binary vector approach, utilizing two plasmids to achieve transformation; the helper plasmid, a disarmed Ri plasmid that provides the vir genes but lacks the oncogenic genes, and the binary vector that contains the gene(s) of interest [6].

Hans-Wilhelm Nützmann and Ramesha Thimmappa (eds.), *Metabolic Gene Clusters: Methods and Protocols*, Methods in Molecular Biology, vol. 3054, https://doi.org/10.1007/978-1-0716-5384-5_11,

The transformation capabilities of *R. rhizogenes*, in conjugation with the binary vector system, enable the generation of transgenic plants, which can be utilized for gene function studies through methods such as silencing or gene editing [5, 7], and the production of proteins and specialized metabolites. As such, hairy root-mediated transformation serves as a powerful tool for studying root biology [8] and plant-microbe interactions [9–11], while the hairy root culture technology has facilitated the isolation of various secondary metabolites, particularly those produced in roots, such as isoflavones from *Trifolium pratense* [12] or steroidal saponins from *Trigonella* species [13, 14]. In this chapter, we describe a detailed protocol for transforming *Lotus japonicus* through an *R. rhizogenes*-mediated hairy root transformation system. The protocol covers the entire process from preparation of the plants and bacterial strains to the methods used to verify the success of the transformation. We emphasize key elements essential for success that can be adapted for use in other legumes. Table 1 provides a comprehensive list of studies employing hairy root transformation of various legume species.

2 Materials

2.1 Seed Scarification, Sterilization, and Germination

1. *Lotus japonicus* seeds (*see* **Note 1**).
2. Sulfuric acid 96% H_2SO_4.
3. Autoclaved deionized H_2O (dH_2O).
4. Sterilization solution: 20% commercial bleach (containing 5% NaOCl), 0.1% Tween 20, 79.9% dH_2O.
5. 0.8% w/v water agar plates as germination medium (*see* **Note 2**): Add 4 g bacteriological agar in 500 mL of dH_2O. Mix well and autoclave at 121 °C for 20 min. Aliquot the medium in round (~20 plates) or square (~10 plates) petri dishes in a laminar flow cabinet. After solidifying, store plates at 4 °C.
6. Glass Pasteur pipettes.
7. Pasteur pipette rubber bulbs.

2.2 R. rhizogenes Culture Preparation

1. Yeast Mannitol Broth (YMB) liquid medium: Add 2 g mannitol, 0.1 g K_2HPO_4, 0.04 g $MgSO_4$ $7H_2O$, 0.02 g NaCl, and 0.2 g Yeast extract in 200 mL of dH_2O. Mix well and aliquot 5 mL of medium into 50-mL Falcon tubes (~40 tubes). Autoclave at 121 °C for 20 min and store at RT.
2. Yeast Mannitol Broth (YMB) solid medium: Add 8 g mannitol, 0.4 g K_2HPO_4, 0.16 g $MgSO_4$ $7H_2O$, 0.08 g NaCl, 0.8 g Yeast extract, and 12 g bacteriological agar in 800 mL of dH_2O. Mix well and autoclave at 121 °C for 20 min. Wait for the agar medium to cool down (~50 °C) and add appropriate antibiotics

Table 1
Hairy root transformation protocols used for legume species

Plant species	Developmental stage	Type of explant and/or preparation	Inoculation method	Strain	References
Alhagi pseudoalhagi	10–15 days old seedlings	Cotyledon and hypocotyl segments	Immersion into suspension	A4	[15]
Arachis hypogaea	Mature seeds	1/3 of embryonic axis	Immersion into suspension	MAFF-02-10266	[16]
Arachis hypogaea	6-week-old plants	Detached leaves	Coated with bacterial paste	K599	[17]
*Arachis hypogaea**	Mature seeds	Embryonic axis	Co-cultivation with *Rhizobium*	R1000	[18]
Astragalus sinicus	5-day-old seedlings	Excision in hypocotyl region	Immersion into suspension	DC-AR2	[19]
Cicer arietinum	5–7 days old seedlings	Radicle excision near collar region	Immersion into suspension	K599	[20]
Cicer arietinum	5–7 days old seedlings	Whole seedling	Injection into cotyledonary node	K599	[20]
Cicer arietinum	–	Radicle excision	Coated with bacterial paste	R1000 ARqua1 MSU440	[10]
*Cicer arietinum**	–	Whole seedlings	Injection into cotyledonary node	R1000 ARqua1 MSU440	[10]
*Glycine max**	7 days old seedlings	Excision in hypocotyl region	Immersion into suspension	K599	[21]
Glycine max	1-day-old seedlings	1/3 part of cotyledon	Immersion into suspension	K599	[5]
Glycine max	1-day-old seedlings	Excision in hypocotyl	Immersion into suspension	K599	[8]
Glycine max	5–10 days old seedlings	Excision in hypocotyl	Immersion into suspension	R1000 A1 ARqua1 K599	[22]
*Glycine max**	5-day-old seedlings	Whole seedling	Coated with bacterial paste	R1000 A1 ARqua1 K599	[22]

(continued)

Table 1
(continued)

Plant species	Developmental stage	Type of explant and/or preparation	Inoculation method	Strain	References
Glycine max	7–10 days old seedlings	Whole seedling	Injection into cotyledonary node	K599	[23]
Glycine max	18–20 days old plants	Detached leaves	Coated with bacterial paste	K599	[23]
*Glycine max**	5-day-old seedlings	Whole seedling	1. Immersion into suspension 2. Coated with bacterial paste	K599	[3]
Glycine max	5–7 days old seedlings	Excised cotyledons	Immersion into suspension with wounded cotyledons	K599	[24]
Glycine max	4-day-old seedlings	Excised shoots	Coated with bacterial paste	K599	[25]
Lens culinaris	3-day-old seedlings	Excision in hypocotyl	Immersion into suspension	K599	[26]
Lens culinaris	3-day-old seedlings	Excision in hypocotyl region	Immersion into suspension	K599 R1000 ARqual	[27]
Lotus corniculatus	Superroot culture	Excised shoots	Immersion into suspension	K599	[28]
Lotus japonicus	6-day-old seedling	Radicle excision	Immersion into suspension	MSU440	[29]
Lotus japonicus	4–5 days old seedling	Excision in hypocotyl	Immersion into suspension	–	[30]
Lotus japonicus	2-day-old seedling	Radicle excision	Coated with bacterial paste	LBA 9402	[31]
Lotus japonicus	3-day-old seedlings	Whole seedling	Injection into cotyledonary node	AR1193 C58CI	[32]
Lotus japonicus	5-day-old seedling	Whole seedling	Injection into hypocotyl area	15834 AR10 MSU440 ARqual LBA 9402 K599 LBA 4404 8196	[33]

(continued)

Table 1
(continued)

Plant species	Developmental stage	Type of explant and/or preparation	Inoculation method	Strain	References
Lotus japonicus	7-day-old seedling	Radicle excision	Immersion into suspension	LBA 1334 K599 LBA 9402	[34]
Medicago truncatula	2-day-old seedlings	Radicle excision	Coated with bacterial paste	ARqua1	[35, 36]
Medicago truncatula	–	Detached leaves	Co-cultivation	LBA 9402	[37]
Phaseolus vulgaris	3-day-old seedlings	Whole seedling	Injection into cotyledon node	K599 R1000	[38]
Phaseolus vulgaris	3-day-old seedlings	Excision under the cotyledon	Immersion into suspension	K599 R1000	[38]
Phaseolus vulgaris	3-day-old seedling	Whole seedling	Wound hypocotyl area and coat with bacterial paste	K599	[39]
*Phaseolus vulgaris**	5-day-old seedlings	Whole seedling	Injection into cotyledonary node	K599	[40, 41]
Phaseolus vulgaris	–	Detached leaves	Coated with bacterial paste	K599	[42]
Phaseolus vulgaris	3–8 days old seedlings	Excision in hypocotyl	Immersion into suspension	K599	[43]
*Phaseolus vulgaris**	3–8 days old seedlings	Whole seedling	Injection into cotyledonary node	K599	[43]
Pisum sativum	8-day-old seedlings	Excision in hypocotyl	Immersion into suspension	ARqua1 AR1193	[44]
Pisum sativum	3-day-old seedlings	Excision in hypocotyl	Immersion into suspension	K599	[26]
*Pisum sativum**	10-day-old plants	Excision in epicotyl	Co-cultivation with *Rhizobium*	AR1193	[9]
Sesbania rostrata	2-day-old seedlings	Whole seedling	Immersion into suspension	15834 2659 8196 1724 C58C1	[45]
Trifolium pratensis	4-day-old	Whole seedling	Injection into stem area	LBA 1334	[46]
Trifolium repens	6–8 days old seedlings	Whole seedling	Injection into stem area	LBA 1334	[47]

(continued)

Table 1
(continued)

Plant species	Developmental stage	Type of explant and/or preparation	Inoculation method	Strain	References
Trigonella foenum graecum	4-day-old seedling	Radicle excision	1. Immersion in suspension 2. Coated with bacterial paste	ARqual LBA 9402	[48, 49]
Trigonella foenum graecum	4–5 weeks old plants	Detached leaves Leaf cotyledons Hypocotyl	Immersion into suspension	15834 Wt A4	[14]
Trigonella monantha	4–5 weeks old plants	Detached leaves Leaf cotyledons Hypocotyl	Immersion into suspension	15834 Wt A4	[14]
Vicia sativa	5-day-old seedlings	Whole seedling	Coated with bacterial paste	K599	[50]
Vicia sativa	5-day-old seedlings	Excised cotyledons	Coated with bacterial paste	K599	[50]
*Vigna unguiculata**	12–14 days old plants	Excision below cotyledon	Immersion into suspension	R1000	[51]

Basic principle for the developed protocols remains the same while key factors towards adapting and optimizing a transformation protocol fit-for-species include the developmental stage of the plant, the type of explant, the strain of R. rhizogenes, and the inoculation method. Most protocols or studies describe the use of in vitro hairy root transformation, but there are a few cases where the ex vitro methodology is employed, as indicated by an (*)

for *R. rhizogenes* selection (*see* **Note 3**). Aliquot the medium in round (~30 plates) petri dishes in a laminar flow cabinet. After solidifying, store plates at 4 °C.

3. Rifampicin stock solution (25 mg/mL, DMSO) (*see* **Note 4**).
4. Micropulser electroporator machine (Biorad).
5. Electroporation cuvettes (Biorad).
6. *R. rhizogenes* (LBA 9402) electro-competent cells (*see* **Notes 5** and **6**).
7. 0.22 μm filters.
8. Liquid nitrogen.
9. 50% (v/v) glycerol (diluted in dH_2O and autoclaved at 121 °C for 20 min).

2.3 Hairy Root Transformation and Regeneration

1. ½ MS agar plates as co-cultivation medium (*see* **Note 7**): Add 2.2 g Murashige & Skoug salts including vitamins (Duchefa Biochemie, Belgium) and 30 g sucrose in 1 L of dH_2O and mix well (*see* **Notes 8** and **9**). Adjust pH to 5.6 ± 0.1 with 1 M KOH. Add 8 g bacteriological agar and autoclave at 121 °C for 20 min. Aliquot the medium in square (~20 plates) or round (~40 plates) petri dishes in a laminar flow cabinet. After solidifying, store plates at 4 °C.
2. ½ MS agar plates with antibiotic as emergence medium: Add 2.2 g Murashige & Skoug salts including vitamins (Duchefa Biochemie, Belgium) and 30 g sucrose in 1 L of dH_2O and mix well. Adjust pH to 5.6 ± 0.1 with 1 M KOH. Add 8 g bacteriological agar and autoclave at 121 °C for 20 min. Wait for the agar medium to cool down (~50 °C) and add 1 mL 300 mg/mL ceftazidime. Aliquot the medium in square (~20 plates) petri dishes in a laminar flow cabinet. After solidifying, store plates at 4 °C.
3. 0.22 μm filters.
4. 1 M KOH for pH adjustment.
5. Ceftazidime stock solution (300 mg/mL, dH_2O).
6. Sterile deionized H_2O (dH_2O).
7. Filter paper.
8. Sterile forceps.
9. Sterile scalpel bases.
10. Sterile scalpels.

2.4 General Materials and Equipment

Apart from the materials mentioned in Subheadings 2.1, 2.2, and 2.3, the researcher needs to have access to standard materials and equipment used in any molecular biology and plant biotechnology laboratory. For example:

1. Round petri dishes (12 cm diameter).
2. Square petri dishes (10 × 10 cm).
3. 2-mL Eppendorf tubes.
4. 1.5-mL Eppendorf tubes.
5. 50-mL Falcon tubes.
6. Aluminum foil.
7. Parafilm.
8. 3 M Micropore tape.
9. Blowtorch.
10. Growth chamber adjusted to 16 h light/8 h dark photoperiod, 23 ± 2 °C (or other conditions appropriate to the plant of interest).

11. Fume hood.
12. Laminar flow cabinet.
13. Ultra-Low Temperature freezer (−80 °C).
14. Refrigerator (4 °C).
15. Incubator (28 °C).
16. Shaking incubator (28 °C, 160 rpm).
17. Autoclave equipment.
18. Stereoscope equipped with laser and suitable filters for fluorescence detection, or other similar system capable of fluorescence marker detection.
19. Centrifuge.
20. Vortex mixer.
21. PCR thermocycler.
22. Gel electrophoresis apparatus.

3 Methods

3.1 Seed Scarification and Surface Sterilization

An overview of the process starting from Day 1 to Day 8 can be found in Fig. 1.

Day 1

1. Place seeds into sterile 2-mL Eppendorf tube (~100 seeds per tube) (Fig. 3a).

   Next steps take place in a fume hood.
2. Add ~1.8 mL of sulfuric acid (96% H_2SO_4) using a glass Pasteur pipette. All seeds should be covered with sulfuric acid (*see* **Notes 10** and **11**).
3. Incubate seeds for 20 min at room temperature with gentle shaking. As incubation progresses, seeds should start to show signs of discoloration (from dark brown to light brown) (*see* **Notes 12** and **13**).
4. Discard H_2SO_4 carefully using a glass Pasteur pipette (*see* **Note 14**).
5. Wash seeds by adding sterilized cold dH_2O into the tubes, shake them, and discard the water.
6. Repeat 4–5 times to remove any traces of sulfuric acid.
7. Prepare fresh sterilization solution (*see* **Notes 15** and **16**).
8. Add 2 mL of sterilization solution to the tube containing the seeds (*see* **Note 17**).
9. Incubate the seeds for 20 min at room temperature with gentle shaking. As incubation progresses, seeds start turning white (*see* **Note 18**).

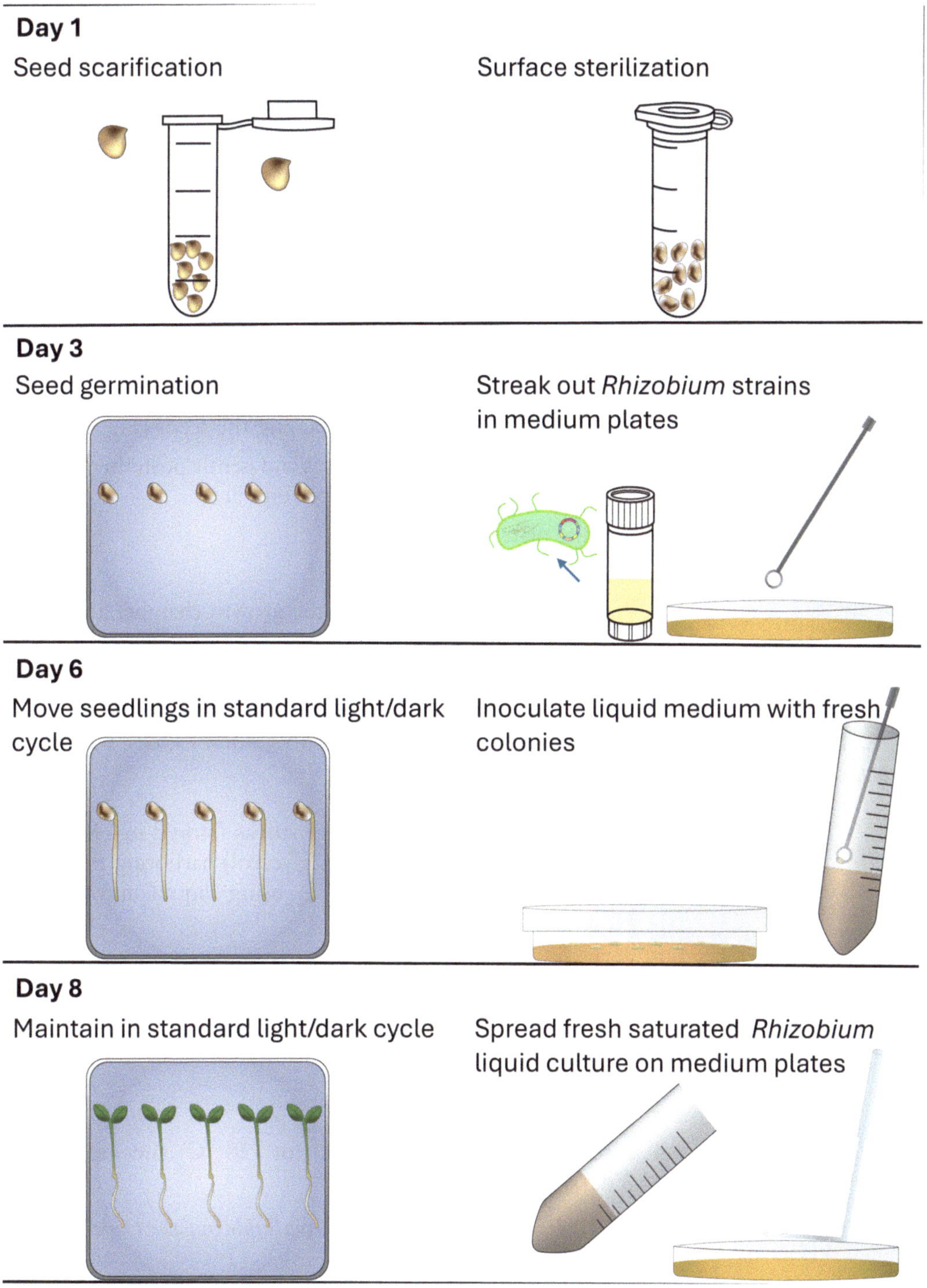

Fig. 1 Schematic representation of the timeline required for the preparation of plants and bacterial strains for hairy root transformation protocol. *See* Subheading 3, Methods for more detailed instructions

Next steps take place in a laminar flow cabinet. Caution must be taken to maintain aseptic conditions.

10. Discard the sterilization solution (*see* **Note 19**).
11. Wash seeds by adding sterilized cold dH_2O into the tubes, shake them, and discard the water.
12. Repeat 5–7 times to remove traces of sterilization solution.
13. Add 2 mL of sterilized dH_2O into the tube. Cover tubes with aluminum foil and keep at 4 °C for 1–2 days (Fig. 3b).

3.2 Seed Germination

Day 3

Next steps take place in a laminar flow cabinet. Extra caution must be taken to maintain aseptic conditions (*see* **Note 20**).

1. Discard water from tubes containing the seeds (Fig. 3c, d).
2. Place seeds into water agar plates and gently push them into the medium using sterilized forceps (Fig. 3e).
3. Seal the plates with parafilm.
4. Cover with aluminum foil.
5. Place the plates at an angle in the growth chamber at 23 °C.
6. Remove aluminum foil after 3 days (Fig. 3f) and maintain plates in an inclined position in the growth chamber (Fig. 3h).

3.3 R. rhizogenes Inoculum Preparation

A. *If R. rhizogenes Strain with Desired Binary Vector is Available*

Day 3

1. Take out *R. rhizogenes* glycerol stocks (stored in YMB liquid medium containing 25% (v/v) glycerol) harboring the desired binary vectors from −80 °C using liquid nitrogen (*see* **Note 21**).

Next steps take place in a laminar flow cabinet (*see* **Note 22**).

2. Streak a small amount of the *R. rhizogenes* onto a YMB round plate supplemented with 25 μg/mL rifampicin and/or other antibiotics for selection.
3. Seal plates with parafilm.
4. Place the plates into an incubator (28 °C) and incubate for 3 days.

B. *If R. rhizogenes Strain is Not Available but the Desired Binary Vector is Verified*

Next steps take place on bench under a blowtorch.

5. Place sterilized electroporation cuvettes on ice.
6. Dilute the desired plasmid vector to a final concentration of 10 ng/μL.

7. Take out *R. rhizogenes* electrocompetent cells from −80 °C and keep them on ice to allow gradual thawing.
8. Set up the machine with the appropriate settings for the electroporation of *Rhizobium*.
9. Add 1 μL of the diluted vector to a 1.5-mL Eppendorf tube of 80–100 μL of *R. rhizogenes* electro-competent cells and mix gently by flicking the tube.
10. Transfer the mixture to a sterilized cold cuvette and deposit the mixture to the bottom by hitting the cuvette onto the bench, making sure that no bubbles remain in the mixture. Wipe the cuvette surface with a napkin to remove any water drops.
11. Insert the cuvette into the cuvette chamber of the micropulser electroporator and deliver a pulse.
12. After the electroporation pulse is delivered, add 1 mL of liquid YMB into the cuvette, mix by pipetting up and down to gently resuspend the cells and transfer resuspended cells into the original 1.5-mL Eppendorf tube.
13. Allow cells to recover by incubating them at 28 °C (with or without shaking) for a minimum of 2 h.
14. Spread ~200 μL of suspended cells onto YMB plates supplemented with 25 μg/mL rifampicin and/or other antibiotics for selection and seal plates with parafilm.
15. Place the plates into an incubator (28 °C) and incubate for 3 days.

Next steps remain the same for both A and B.
Day 6

16. Remove plates from the incubator (Fig. 3g).

Next steps take place in a laminar flow cabinet or on bench under a blowtorch.

17. Using a pipette tip, scrap a single colony off the plates and place the tip with the cells into a sterile 50-mL Falcon tube (falcons), containing 5 mL of liquid YMB supplemented with 25 μg/mL rifampicin and/or other antibiotics for selection (*see* **Note 23**).
18. Transfer falcons into a shaking incubator (28 °C, 160 rpm) and incubate for 2 days.

Day 8

19. Remove falcons with *R. rhizogenes* cultures from incubators.

Next steps take place in a laminar flow cabinet or on bench under a blowtorch.

20. Spread 1 mL of *R. rhizogenes* liquid culture onto YMB round plates supplemented with 25 μg/mL rifampicin and/or other

antibiotics for selection and seal the plates with parafilm. After incubation, a layer of bacteria (bacterial slurry) will be formed (*see* **Note 24**).

21. Transfer plates into an incubator (28 °C) for 2 days (*see* **Note 25**).

3.4 Rhizobium rhizogenes*-Mediated Hairy Root Transformation and Root Regeneration*

An overview of the process starting from Day 10 to >Day 24 can be found in Fig. 2.

Day 10

1. Remove plates with bacterial slurry from incubator (Fig. 3i).

   Next steps take place in a laminar flow cabinet. Extra caution must be taken to maintain aseptic conditions as explained in **Note 20**.

2. Add 3 mL of sterile dH_2O into each plate with the bacterial slurry (*see* **Notes 26** and **27**).
3. Resuspend the cells by scrapping the surface using a pipette tip.
4. Transfer 3 mL of the resuspended cells into a petri dish with a sterilized filter paper.
5. Place a batch (5–10) of 7-day-old seedlings on top of the pre-wet filter paper (*see* **Notes 28** and **29**).
6. Excise the root of each seedling by making a slanted cut right under the hypocotyl using a sterilized scalpel (Fig. 3j, k).
7. Pour a small amount of the resuspended cells into the wounded site and, using sterilized forceps, gently squeeze the wounded end of the hypocotyl (*see* **Note 30**).
8. Transfer the explants into ½ MS co-cultivation medium plates and gently push the wounded site of each explant into the medium. Position 10–20 explants per square plate and 5–10 explants per round plate (*see* **Note 31**).
9. Seal the plates by covering 3 sides with micropore tape and the fourth (bottom) side with parafilm.
10. Cover with aluminum foil and place the plates at an angle in the growth chamber at 23 °C (*see* **Note 32**).
11. Remove aluminum foil after 2 days and maintain plates in a tilted position in the growth chamber for another 3 days (*see* **Note 33**). From now on, maintain plates in a 16 h light/8 h dark photo period.

    Day 15

    Next steps take place in a laminar flow cabinet. Extra caution must be taken to maintain aseptic conditions as explained in **Note 20**.

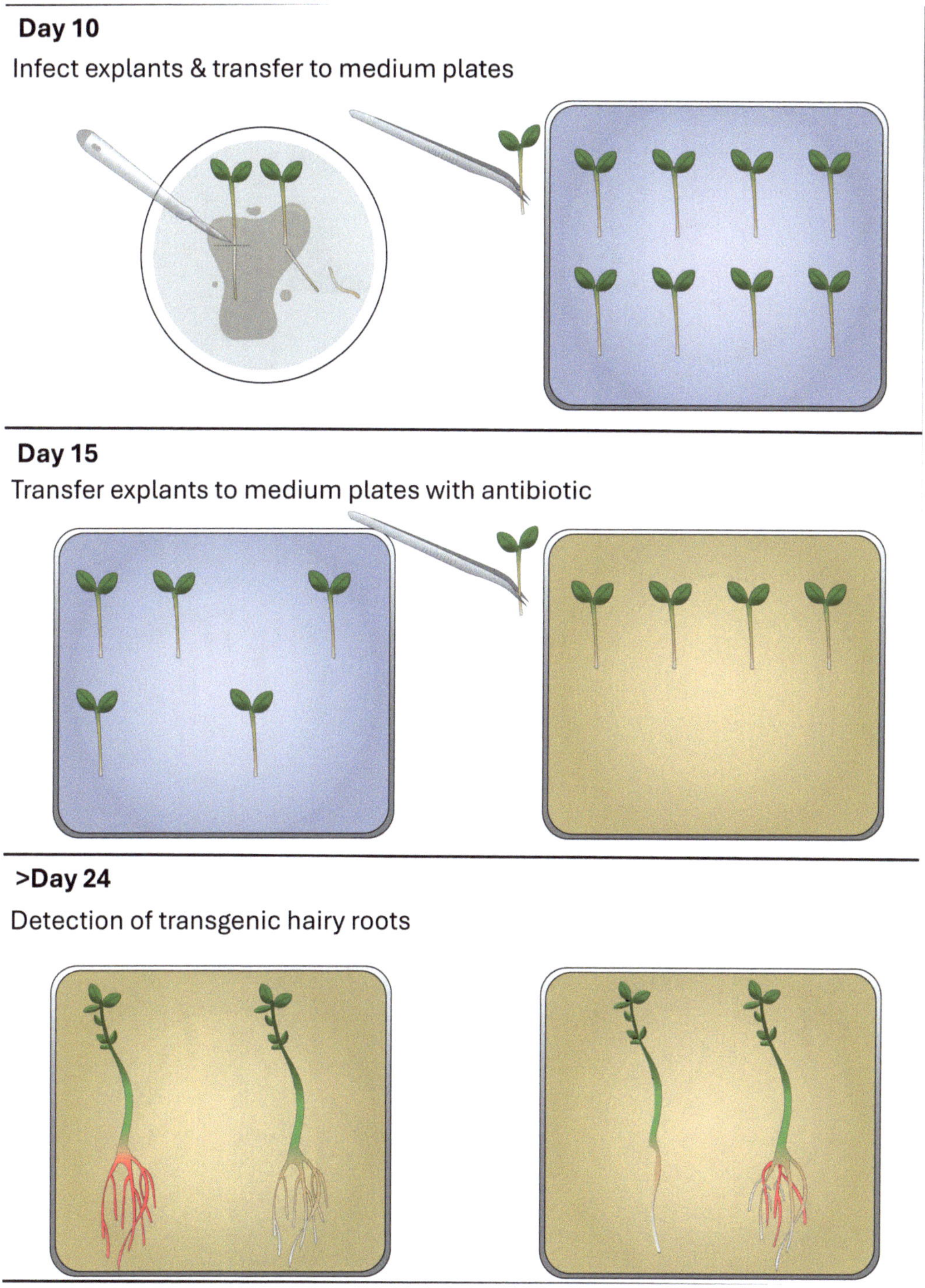

Fig. 2 Schematic representation of the timeline required for the hairy root transformation, root regeneration, and transgene detection. *See* Subheading 3, Methods for more detailed instructions

Fig. 3 Stages during the hairy root transformation of *Lotus japonicus*. (**a**) Seeds before scarification (Day 1). (**b**) Seeds after scarification & surface sterilization (Day 1). (**c**) Seeds after imbibition (Day 3). (**d**) Comparison of seeds before (Day 1) and after (Day 3) imbibition. (**e**) Seeds transferred into medium plates (Day 3). (**f**) Seedlings after 3 days of germinating in the dark (Day 6). (**g**) *R. rhizogenes* growth on a solid medium plate 3 days post streaking (Day 6). (**h**) Seedlings 2 days after being placed to a normal light-dark cycle (Day 8). (**i**) *R. rhizogenes* bacterial slurry (Day 10). (**j**) Scalpel pointing to the site to be infected (Day 10). (**k**) Seedlings with

12. Transfer explants from ½ MS co-cultivation medium plates to ½ MS emergence medium plates (10 explants per plate) and gently push the wounded site of each explant into the medium (*see* **Note 34**) (Fig. 3l).
13. Seal the plates using micropore tape to cover 3 sides and parafilm to cover the fourth (bottom) side.
14. Place the plates at an angle in the growth chamber in a 16 h light/8 h dark photo period at 23 °C until root regeneration (*see* **Note 35**).

3.5 Detection of Transgenic Hairy Roots

Day 24

1. Transfer composite plants from plates to a container filled with 1% bleach for 1 min (*see* **Notes 36** and **37**).
2. Wash plants by submerging them in containers filled with sterilized dH_2O.
3. Repeat 2–3 times to remove traces of bleach.

A. *Lack of Reporter Gene in the T-DNA of the Binary Vector*

4. Depending on downstream application, for every emerging root, either excise the whole root or a segment of it.
5. Isolate DNA from root samples using preferred procedure.
6. Perform a diagnostic PCR, targeting a region within the T-DNA region. Make sure that the produced amplicon is specific for the T-DNA (*see* **Note 38**).
7. Assess results using gel electrophoresis.
8. If the goal is to utilize the composite plants as whole plants (e.g., for functional analysis), it is preferred to remove non-transformed roots with a scalpel.
9. Continue with downstream application.

B. *Inclusion of a Reporter Gene in the T-DNA of the Binary Vector*

10. If the reporter gene encodes for a fluorescent protein (e.g., GFP, mCherry, YFP), use a stereoscope equipped with laser and appropriate filters to assess fluorescence signal in the hairy roots. If the reporter gene encodes for a protein that is involved

Fig. 3 (continued) unfolded cotyledons infected by cutting into the hypocotyl region (Day 10). (**l**) Explants after transferring to medium plates with ceftazidime (Day 15). (**m**) Hairy roots emerging from infected sites in brightfield and through a fluorescence filter to detect tRFP expression (>Day 24). Transformed roots are indicated with the white arrowheads, while non-transformed with yellow arrowheads. The shoot and non-transformed roots are highly visible due to fluorescence overexposure. (**n**) Hairy roots emerging from infected sites 21 days post infection, the RUBY cassette allowing the detection of the transformed roots without specialized equipment (>Day 24)

in the production of a chromophore (i.e., reporter cassette RUBY converting tyrosine to red betalain), assessment can be made without the use of any special equipment (*see* **Note 39**) (Fig. 3m, n).

11. If the goal is to utilize the composite plants as whole plants, it is preferred to remove non-transformed roots with a scalpel.
12. Continue with downstream application.

4 Notes

1. Seed age, storage conditions of seeds, and stage at which seeds were collected are some of the factors affecting the germination rate. A higher percentage of germinated seeds can be achieved by selecting newer seeds that have been collected after reaching maturity and stored in a dry and dark place.
2. Instead of water agar plates, round plates with two layers of pre-wet filter paper can be used as germination medium.
3. Antibiotics are added to the medium post-autoclaving as they are inactivated by high temperatures.
4. Antibiotic stock solutions are prepared at 1000x the concentration required in the medium and stored appropriately. For example, rifampicin and ceftazidime stocks were prepared at concentrations of 25 mg/mL and 300 mg/mL, respectively, and stored at 4 °C, while the concentrations required in the medium were 25 μg/mL and 300 μg/mL. To minimize risk of contamination, antibiotic stock solutions are filtered through a 0.22 μm filter.
5. The transformation efficiency is greatly influenced by the selection of the *Rhizobium* strain [22, 27]. Selecting a suitable strain, which is compatible with the plant species to be transformed, is important to ensure a sufficiently high transformation rate. Apart from the LBA 9402 stain, the LBA 1334 strain is also capable of transforming *Lotus japonicus* [34].
6. The preparation of *Rhizobium rhizogenes* electro-competent cells is not described here.
7. All media should be made fresh to reduce the risk of contamination, particularly those for the cultivation of explants or seedlings.
8. The basis of most "co-cultivation and emergence" media employed for hairy root transformation is basal media. Most of them are based on MS: Murashige & Skoog [27, 28, 38], Gamborg B5 [24, 30], and B&D: Broughton and Dilworth [8] media. In many cases, apart from the basal medium, vitamins [24, 37], growth regulators [11], and acetosyringone [AS] [10, 31, 38] are included to increase transformation efficiency and promote root growth.

9. Adding sucrose to the co-cultivation and emergence medium was found to increase the transformation efficiency [1] and promote faster root growth [48].
10. Scarification, in this case by incubating seeds in sulfuric acid, is a process used to soften the seed coat and help promote water uptake and germination. This step can be omitted when dealing with plant species without hard seed coats.
11. When managing sulfuric acid, take extra care by wearing protective gear because of its corrosive properties. Wear double gloves, a lab coat, and protective glasses.
12. Incubation period in sulfuric acid depends on seed age and genotype (i.e., wild type or mutant). Some mutant and older seeds usually require less than 20 min (e.g., 12–15 min).
13. During incubation, seeds tend to clump together or stick to the tube walls. To ensure maximum contact with either sulfuric acid or sterilization solution, make sure to separate the seeds periodically by shaking the tubes.
14. When discarding sulfuric acid, use a container filled with cold water to dilute it.
15. Seeds are sterilized to remove surface contaminants to maintain aseptic conditions in the subsequent in-vitro assay.
16. In the sterilization solution, bleach is used as disinfectant, while Tween 20 serves as a wetting agent, improving water uptake by reducing surface tension.
17. Before adding the sterilization solution, seeds can be washed with 70% ethanol for 5 min as an additional step to minimize contamination.
18. Sterilization period also depends on seed age and genotype. Mutant and older seeds usually require less than 20 min (e.g., 12–15 min).
19. When incubating with the sterilization solution, move to the washing promptly as prolonged exposure of the seeds to bleach may impede germination.
20. Before any operation, especially one involving plant handling, the inside of the laminar flow cabinet should be sterilized with UV light, followed by spraying the laminar bench with 70% ethanol and wiping it clean. Petri dishes, pipette tip containers, pipettes, and any other objects entering the laminar after UV sterilization must be wiped clean with 70% ethanol. Hands are also sprayed with 70% ethanol to minimize skin-residing microbes. Forceps and scalpel bases used for plant handling should be autoclaved at 121 °C for 20 min before use, scalpels must be also sterilized, while forceps and scalpels mounted to scalpel bases are repeatedly sterilized during the germination

stage, hairy root transformation and subsequent transfer to other mediums by dipping the ends in 70% ethanol solution and holding them in the flame for a few seconds (~5 s).

21. *R. rhizogenes* plates cannot be maintained at 4 °C for long, so fresh plates must be prepared for each round of hairy root transformation.
22. Alternatively, streaking the *R. rhizogenes* strains from glycerol stocks can be performed on the bench, under a blowtorch to maintain aseptic conditions.
23. A 5 mL YMB bacterial liquid culture can be used for preparing up to five YMB bacterial slurry agar plates.
24. A single bacterial slurry plate can be used to transform 30–40 seedlings. So, depending on the number of seedlings needed to be transformed per binary vector, an appropriate number of bacterial slurry plates is prepared.
25. Normally, YMB plates with bacteria are placed lid-side down to prevent contamination. However, when preparing the bacterial slurry, plates are placed lid-side up, as the amount of liquid culture added to the agar plate will not be absorbed.
26. Instead of using sterile dH_2O to dilute the bacterial slurry, liquid ½ MS co-cultivation medium may be used.
27. The addition of acetosyringone enhances the transformation efficiency in many legumes [22]. In most cases, acetosyringone is added to the co-cultivation medium during agar plate preparation [27] or to the solution used to resuspend the cells [22].
28. Seedling age is one of the factors influencing the transformation efficiency. Using older seedlings for hairy root transformation may negatively affect the efficiency, as shown in some studies [21, 40].
29. Select healthy seedlings with unfolded green cotyledons and well-developed roots to increase the chances of survival during co-inoculation with *R. rhizogenes* and hairy root regeneration.
30. There are various alternatives to the inoculation method described in this protocol. Instead of resuspending the cells of the bacterial slurry, bacterial liquid cultures can be directly used, either as they are (*see* Day 6 of Subheading 3.4 of Methods), diluted with sterilized dH_2O or subjected to centrifugation to collect the cells and resuspending them in either sterilized dH_2O or another co-inoculation medium (e.g., ½ MS basal medium). In addition, the OD_{600} of the bacterial suspension is also taken into consideration, with most studies suggesting $OD_{600} = 0.6$ as the optimal optical density [15, 20, 21]. Another approach, based on the use of bacterial slurry, includes coating the cut ends of the hypocotyl regions with a thick layer of bacterial paste by dipping them into the slurry or

deposing some of it onto the cut end. Alternatively, instead of excising the root, inoculation is possible by wounding the plant in the hypocotyl region and placing it in a bacterial slurry/ suspension or by injecting the bacterial suspension directly into the plant.

31. Despite the efforts of maintaining aseptic conditions, fungal contamination may appear on the plates. In those cases, transfer plants distant to the contamination site into new plates.
32. The transformation efficiency is also affected by the temperature of the co-inoculation stage [3, 40]. *Lotus japonicus* explants are inoculated with Rhizobia at 23 °C ± 2 °C, but, depending on the legume species, values can vary between 22 and 28 °C.
33. The co-inoculation period is one of the most important factors affecting transformation efficiency. In the case of *Lotus japonicus*, incubating explants with *R. rhizogenes* in a nutrient-rich medium (i.e., ½ MS, 3% w/v sucrose) for 5 days is a standard practice, but for other species it varies according to explant type and growth medium.
34. Sometimes growth of *R. rhizogenes* is observed in ½ MS plates supplemented with ceftazidime. As explant survival and root regeneration is aborted by bacterial overgrowth, it is strongly recommended to transfer plants to new emergence medium plates at the first sign of Rhizobium growth.
35. Hairy roots appear around 2 weeks post inoculation with *R. rhizogenes*. Roots emerging at an earlier time point, especially away from the infection site, are non-transformed and can be excised to facilitate subsequent handling of emerging transformed ones.
36. Plants exhibiting a hairy root phenotype are the ones to have most likely been transformed with both the T-DNA of the Ri plasmid and the T-DNA of the binary vector. Plants with hairy root phenotype are recognized by having multiple roots emerging from the infected site and exhibiting a lack of geotropism. Each emerging root within a hairy root system is considered a separate transformation event and, as a result, some roots may not be transformed.
37. Plants are sterilized with 1% bleach to remove rhizobia.
38. When employing a diagnostic PCR to detect the presence of the T-DNA region in roots, also amplify for non T-DNA regions of the vector to ascertain that no false positive transgenic plants are detected due to the presence of left-over *Rhizobium*.

39. Incorporating a reporter gene in the T-DNA of the binary vector is recommended since it allows for visual detection of transformed roots without damaging the plants or altering the root structure.

Acknowledgments

We thank Dimitra Vasilakopoulou for her contribution to the graphical figures.

References

1. Bandaranayake PCG, Yoder JI (2018) Factors affecting the efficiency of Rhizobium rhizogenes root transformation of the root parasitic plant Triphysaria versicolor and its host Arabidopsis thaliana. Plant Methods 14:61
2. Winans SC (1992) Two-way chemical signaling in Agrobacterium-plant interactions. Microbiol Rev 56:12–31
3. Kereszt A, Li D, Indrasumunar A et al (2007) Agrobacterium rhizogenes-mediated transformation of soybean to study root biology. Nat Protoc 2:948–952
4. Gantait S, Mukherjee E (2021) Hairy root culture technology: applications, constraints and prospect. Appl Microbiol Biotechnol 105:35–53
5. Cheng Y, Wang X, Cao L et al (2021) Highly efficient Agrobacterium rhizogenes-mediated hairy root transformation for gene functional and gene editing analysis in soybean. Plant Methods 17:73
6. Bahramnejad B, Naji M, Bose R et al (2019) A critical review on use of Agrobacterium rhizogenes and their associated binary vectors for plant transformation. Biotechnol Adv 37: 107405
7. Cai Y, Chen L, Liu X et al (2015) CRISPR/Cas9-mediated genome editing in soybean hairy roots. PLoS One 10:e0136064
8. Huang P, Lu M, Li X et al (2022) An efficient Agrobacterium rhizogenes-mediated hairy root transformation method in a soybean root biology study. Int J Mol Sci 23:12261
9. Clemow SR, Clairmont L, Madsen LH et al (2011) Reproducible hairy root transformation and spot-inoculation methods to study root symbioses of pea. Plant Methods 7:46
10. Mandal D, Sinharoy S (2019) A toolbox for nodule development studies in chickpea: a hairy-root transformation protocol and an efficient laboratory strain of *Mesorhizobium* sp. Mol Plant Microbe Interact 32:367–378
11. Iantcheva A, Mysore KS, Ratet P (2013) Transformation of leguminous plants to study symbiotic interactions. Int J Dev Biol 57:577–586
12. Anil Kumar M, Sravanthi Pammi SS, Sukanya MS et al (2018) Enhanced production of pharmaceutically important isoflavones from hairy root rhizoclones of Trifolium pratense L. Vitro Cell Dev Biol Plant 54:94–103
13. Zolfaghari F, Rashidi-Monfared S, Moieni A et al (2020) Improving diosgenin production and its biosynthesis in Trigonella foenum-graecum L. hairy root cultures. Ind Crops Prod 145:112075
14. Kohsari S, Rezayian M, Niknam V et al (2020) Antioxidative enzymes activities and accumulation of steroids in hairy roots of Trigonella. Physiol Mol Biol Plants 26:281–288
15. Wang YM, Wang JB, Luo D et al (2001) Regeneration of plants from callus cultures of roots induced by Agrobacterium rhizogenes on Alhagi pseudoalhagi. Cell Res 11:279–284
16. Akasaka Y (1998) Morphological alterations and root nodule formation in Agrobacterium rhizogenes-mediated transgenic hairy roots of peanut (Arachis hypogaea L.). Ann Bot 81: 355–362
17. Guimaraes LA, Pereira BM, Araujo ACG et al (2017) Ex vitro hairy root induction in detached peanut leaves for plant–nematode interaction studies. Plant Methods 13:25
18. Sinharoy S, Saha S, Chaudhury SR et al (2009) Transformed hairy roots of *Arachis hypogea*: a tool for studying root nodule symbiosis in a non–infection thread legume of the Aeschynomeneae tribe. Mol Plant Microbe Interact 22: 132–142
19. Cho H-J, Widholm JM (2002) Improved shoot regeneration protocol for hairy roots of

the legume Astragalus sinicus. Plant Cell Tissue Organ Cult 69:259–269
20. Aggarwal PR, Nag P, Choudhary P et al (2018) Genotype-independent Agrobacterium rhizogenes-mediated root transformation of chickpea: a rapid and efficient method for reverse genetics studies. Plant Methods 14:55
21. Fan Y, Zhang X, Zhong L et al (2020) One-step generation of composite soybean plants with transgenic roots by Agrobacterium rhizogenes-mediated transformation. BMC Plant Biol 20:208
22. Niazian M, Belzile F, Curtin SJ et al (2023) Optimization of in vitro and ex vitro Agrobacterium rhizogenes-mediated hairy root transformation of soybean for visual screening of transformants using RUBY. Front Plant Sci 14:1207762
23. Freitas-Alves NS, Moreira-Pinto CE, Arraes FBM et al (2024) An ex vitro hairy root system from petioles of detached soybean leaves for in planta screening of target genes and CRISPR strategies associated with nematode bioassays. Planta 259:23
24. Liu J, Gunapati S, Mihelich NT et al (2019) Genome editing in soybean with CRISPR/Cas9. In: Qi Y (ed) Plant genome editing with CRISPR systems. Springer New York, New York, pp 217–234
25. Song J, Tóth K, Montes-Luz B et al (2021) Soybean hairy root transformation: a rapid and highly efficient method. Curr Protoc 1:e195
26. Xiong R, Wang A, Marsolais F et al (2023) A fast and efficient method to introduce apple latent spherical virus to legume plants via Agrobacterium rhizogenes-mediated transformation of hairy roots. Legume Sci 5:e154
27. Foti C, Pavli O (2020) High-efficiency Agrobacterium rhizogenes-mediated transgenic hairy root induction of Lens culinaris. Agronomy 10:1170
28. Jian B, Hou W, Wu C et al (2009) Agrobacterium rhizogenes-mediated transformation of Superroot-derived Lotus corniculatus plants: a valuable tool for functional genomics. BMC Plant Biol 9:78
29. Martirani L, Stiller J, Mirabella R et al (1999) T-DNA tagging of nodulation- and root-related genes in *Lotus japonicus:* expression patterns and potential for promoter trapping and insertional mutagenesis. Mol Plant Microbe Interact 12:275–284
30. Okamoto S, Yoro E, Suzaki T et al (2013) Hairy root transformation in Lotus japonicus. Bio Protoc 3(12). https://doi.org/10.21769/BioProtoc.795
31. Xu Y, Liu F, Han G et al (2018) Improvement of *Lotus japonicus* hairy root induction and development of a mycorrhizal symbiosis system. Appl Plant Sci 6:e1141
32. Ferguson S, Abel NB, Reid D et al (2023) A simple and efficient protocol for generating transgenic hairy roots using Agrobacterium rhizogenes. PLoS One 18:e0291680
33. Stiller J, Martirani L, Tuppale S et al (1997) High frequency transformation and regeneration of transgenic plants in the model legume *Lotus japonicus.* J Exp Bot 48:1357–1365
34. Wang Y, Yang F, Zhu P-F et al (2021) Use of the rhizobial type III effector gene nopP to improve Agrobacterium rhizogenes-mediated transformation of Lotus japonicus. Plant Methods 17:66
35. Boisson-Dernier A, Chabaud M, Garcia F et al (2001) *Agrobacterium rhizogenes*-transformed roots of *Medicago truncatula* for the study of nitrogen-fixing and endomycorrhizal symbiotic associations. Mol Plant Microbe Interact 14:695–700
36. Crane C, Wright E, Dixon RA et al (2006) Transgenic Medicago truncatula plants obtained from Agrobacterium tumefaciens-transformed roots and Agrobacterium rhizogenes-transformed hairy roots. Planta 223:1344–1354
37. Zhang H, Cao Y, Zhang H et al (2020) Efficient generation of CRISPR/Cas9-mediated homozygous/biallelic Medicago truncatula mutants using a hairy root system. Front Plant Sci 11:294
38. Li HL, Fang PP, Hu YN et al (2022) An Agrobacterium rhizogenes strain R1000-mediated efficient hairy root transformation protocol for common bean. Legume Res Int J. https://doi.org/10.18805/LRF-680
39. Pacheco R, Estrada-Navarrete G, Nava N et al (2023) Hairy root generation in common bean (Phaseolus vulgaris L.) and selection of Agrobacterium rhizogenes clones v2. https://www.protocols.io/view/hairy-root-generation-in-common-bean-phaseolus-vul-cu92wz8e
40. Estrada-Navarrete G, Alvarado-Affantranger X, Olivares J-E et al (2007) Fast, efficient and reproducible genetic transformation of Phaseolus spp. by Agrobacterium rhizogenes. Nat Protoc 2:1819–1824
41. Estrada-Navarrete G, Alvarado-Affantranger X, Olivares J-E et al (2006) *Agrobacterium rhizogenes* transformation of the *Phaseolus* spp.: a tool for functional genomics. Mol Plant Microbe Interact 19:1385–1393

42. Liu YH, Cai XQ, Ning K et al (2024) An efficient hairy root transformation method for common bean based on petiole explants. Legume Res Int J. https://doi.org/10.18805/LRF-778
43. Khandual S, Reddy PM (2014) Rapid, efficient and high-performance protocol for Agrobacterium rhizogenes-mediated hairy root transformation of the common bean Phaseolus vulgaris. Adv Biosci Biotechnol 05:333–339
44. Leppyanen IV, Kirienko AN, Dolgikh EA (2019) *Agrobacterium rhizogenes*-mediated transformation of *Pisum sativum* L. roots as a tool for studying the mycorrhizal and root nodule symbioses. PeerJ 7:e6552
45. Van De Velde W, Mergeay J, Holsters M et al (2003) Agrobacterium rhizogenes-mediated transformation of Sesbania rostrata. Plant Sci 165:1281–1288
46. Díaz CL, Spaink HP, Kijne JW (2000) Heterologous rhizobial lipochitin oligosaccharides and chitin oligomers induce cortical cell divisions in red clover roots, transformed with the pea lectin gene. Mol Plant Microbe Interact 13: 268–276
47. Díaz CL, Melchers LS, Hooykaas PJJ et al (1989) Root lectin as a determinant of host–plant specificity in the Rhizobium–legume symbiosis. Nature 338:579–581
48. Garagounis C, Georgopoulou M-E, Beritza K et al (2020) An Agrobacterium rhizogenes mediated hairy root transformation protocol for fenugreek. MethodsX 7:101098
49. Garagounis C, Beritza K, Georgopoulou M-E et al (2020) A hairy-root transformation protocol for Trigonella foenum-graecum L. as a tool for metabolic engineering and specialised metabolite pathway elucidation. Plant Physiol Biochem 154:451–462
50. Nguyen V, Searle IR (2022) An efficient root transformation system for recalcitrant Vicia sativa. Front Plant Sci 12:781014
51. Mellor KE, Hoffman AM, Timko MP (2012) Use of ex vitro composite plants to study the interaction of cowpea (Vigna unguiculata L.) with the root parasitic angiosperm Striga gesnerioides. Plant Methods 8:22

Chapter 12

CUT&Tag Epigenomic Profiling of Biosynthetic Gene Clusters in *Arabidopsis thaliana*

Jessica Taylor, Kieran Reynolds, and Hans-Wilhelm Nützmann

Abstract

Plants synthesise a diverse array of specialised metabolites with important functions in environmental adaptation and reproduction. Biosynthesis genes arranged in genomic clusters typically exhibit shared epigenomic features that likely contribute to their transcriptional co-regulation. Here, we describe a protocol for the genome-wide profiling of histone modifications in *Arabidopsis thaliana* by CUT&Tag. In addition, we provide an overview about the bioinformatic analysis of tissue-specific enrichment patterns of epigenetic marks within biosynthetic gene clusters.

Key words CUT&Tag, Biosynthetic gene cluster, Epigenetics, Histone modification, *Arabidopsis thaliana*

1 Introduction

The survival and adaptation of plants to dynamic environmental conditions rely on the coordination of many complex and diverse metabolic processes. Genes involved in the biosynthesis of specialised metabolic compounds are often arranged in pathway-specific clusters, which may facilitate their co-regulation [1–3]. Indeed, the correlated expression of biosynthetic gene clusters (BGCs) has been linked to correlated changes in chromatin modifications, formation of isolated 3D contact domains, and distinct localisation within the nuclear space. For example, the thalianol gene cluster of *A. thaliana*, which is expressed in roots but silenced in leaves, exhibits marked differences in epigenomic enrichment profiles and 3D organisation [3, 4]. Further investigation into the mechanism of transcriptional co-regulation within BGCs is crucial to our understanding of the evolutionary importance of gene clustering and its manipulation for crop improvement [5].

In this chapter, we outline a method for analysing the histone modifications of BGCs and associated regulatory regions in

Hans-Wilhelm Nützmann and Ramesha Thimmappa (eds.), *Metabolic Gene Clusters: Methods and Protocols*, Methods in Molecular Biology, vol. 3054, https://doi.org/10.1007/978-1-0716-5384-5_12,

Arabidopsis thaliana leaf and root tissue using the CUTANA™ CUT&Tag Kit with a modified protocol. We also provide a step-by-step bioinformatic pipeline for sequencing data processing, alignment, peak-calling, and visualisation.

2 Materials

2.1 Growth Media

1. MS Phytagel (pH 5.8): 0.43% (w/v) Murashige & Skoog Medium, 1% (w/v) sucrose, 0.5% (w/v) phytagel. Pour into 100 mm square Petri dishes immediately after autoclaving.

2.2 Nuclei Isolation

2.2.1 Instruments and Equipment

1. Miracloth.
2. Nylon net filter (30 μm pore size).
3. Axygen™ multi-well low-profile reagent reservoirs.
4. Razor blades.
5. Neubauer counting chamber.
6. Confocal microscope.

2.2.2 Stock Solutions and Reagents

1. Glycerol.
2. 1 M sucrose.
3. 1 M HEPES pH 8.
4. 1 M TritonX-100.
5. 1 M KCl.
6. 1 M $MgCl_2$.
7. 0.1 M PMSF.
8. β-mercaptoethanol.
9. DAPI stain.

2.3 CUT&Tag

2.3.1 Instruments and Equipment

1. Magnetic rack for 1.5 mL Eppendorf tubes.
2. Magnetic rack for 0.1 mL PCR tubes.
3. Nutator.
4. PCR thermocycler.
5. Qubit Fluorometer (Thermo Fisher Scientific).
6. 2100 Bioanalyzer system (Agilent).

2.3.2 Stock Solutions and Reagents

1. CUTANA™ CUT&Tag Kit (Epicypher).
2. 25 X Protease Inhibitor (Roche).
3. 80% Ethanol.
4. Rabbit primary antibodies against histone modifications of choice.

5. Qubit dsDNA HS Assay Kit.
6. Agilent High Sensitivity DNA Kit.

2.3.3 Nuclei Isolation Buffer

1. Nuclei Isolation Buffer: 40% glycerol (v/v), 250 mM sucrose, 20 mM HEPES pH 8, 0.25% TritonX-100 (v/v), 5 mM KCl, 1 mM $MgCl_2$, 0.1 mM PMSF (TOXIC; add under fume hood immediately prior to use), 0.1% β-mercaptoethanol (TOXIC; add under fume hood immediately prior to use). Always prepare fresh and keep on ice throughout the experiment.

2.3.4 CUT&Tag Buffers (Volumes for x1 Sample Plus Excess Volume to Account for Pipetting Error; Always Prepare Fresh and Keep on Ice/4 °C)

1. Nuclei Buffer: 235 μL Pre-Nuclear Extraction Buffer, 9.8 μL 25X Protease Inhibiter, 0.13 μL Spermidine.
2. Wash Buffer 1: 1.3 mL Pre-Wash Buffer, 56 μL 25X Protease Inhibitor, 0.7 μL Spermidine, 2.8 μL Digitonin.
3. Wash Buffer 2: 600 μL Wash Buffer 1, 20.7 μL NaCl.
4. Antibody Buffer: 60 μL Wash Buffer 1, 0.25 μL EDTA.
5. Tagmentation Buffer: 60 μL Wash Buffer 2, 0.6 μL $MgCl_2$.

2.4 Data Analysis

1. Trim Galore (v0.6.7).
2. Bowtie2 (v2.4.4) [6].
3. Picard Toolkit (v3.3.0).
4. samtools (v1.13) [7].
5. deepTools (v3.5.1) [8].
6. csaw (v 1.44.0) [9].
7. Integrative Genome Viewer (IGV, v2.19.1) [10].

3 Methods

3.1 Nuclei Isolation

1. Surface sterilise *A. thaliana* seeds and grow vertically on MS phytagel for 7–10 days under long-day conditions (16 h light at 22 °C/8 h dark at 18 °C).
2. Harvest cotyledons and/or roots using a razor blade and forceps. Dry the plant tissue by gently pressing between layers of Miracloth and tissue paper.
3. Transfer the plant tissue into a multi-well reagent reservoir. Submerge the plant tissue in Nuclei Isolation Buffer, just enough to cover the tissue. Use a razor blade to finely chop for 10 min.
4. Filter the nuclei suspension into a 50 mL Falcon tube through one layer of Miracloth then one layer of nylon net filter with Nuclei Isolation Buffer. For nuclei isolation from root tissue, filter an additional time through a fresh layer of nylon net filter.

5. Centrifuge nuclei suspension at 3000 g, 4 °C, for 15 min. Discard supernatant and resuspend nuclei in 1000 μL Nuclei Isolation Buffer. Transfer to a 1.5 mL Eppendorf tube.
6. Centrifuge nuclei suspension at 1900 g, 4 °C, for 5 min. Discard supernatant and resuspend nuclei in 1000 μL Nuclei Isolation Buffer. For nuclei isolation from root tissue, repeat this step once more.
7. Centrifuge nuclei suspension at 1900 g, 4 °C, for 5 min. Discard supernatant and resuspend nuclei in 500 μL Nuclei Isolation Buffer. Keep on ice.

3.2 Nuclei Counting

1. Start-up the confocal microscope.
2. Aliquot 10 μL nuclei suspension into a fresh 1.5 mL Eppendorf tube. Add DAPI stain and incubate in the dark for 10 min.
3. Load 10 μL of stained nuclei suspension onto the Neubauer counting chamber.
4. Examine nuclei using 20X objective lens with DAPI filter.
5. Calculate the nuclei counts using the following equation:

$$\text{Nuclei per } \mu\text{L} = \frac{\text{Counted nuclei}}{\text{Counted area } (\text{mm}^2) \times \text{Chamber depth } (\text{mm})} \times 500$$

3.3 CUT&Tag Day 1 (Modified from CUTANA™ CUT&Tag Kit 2.0 User Manual)

1. Prepare Nuclei Buffer, Wash Buffer 1, Wash Buffer 2, and Antibody Buffer for n samples plus positive and negative controls. Keep Nuclei Buffer and Antibody Buffer on ice. Store Wash Buffer 1 and Wash Buffer 2 at 4 °C.
2. Transfer 50,000–100,000 nuclei per sample into 1.5 mL Eppendorf tubes.
3. Centrifuge nuclei suspension at 1900 g, 4 °C, for 5 min. Discard supernatant and resuspend in 105 μL Nuclei Buffer using wide-bore tips (*see* **Note 1**). Repeat once more.
4. Resuspend ConA beads by gently pipetting (*see* **Note 2**). Transfer 11 μL per sample into 1.5 mL Eppendorf tubes. Place tube on magnetic rack for 2 min then, once clear, discard supernatant.
5. Remove tube from magnetic rack and resuspend in 100 μL per sample Bead Activation Buffer by gently pipetting. Place tube on magnetic rack for 2 min then, once clear, discard supernatant.
6. Repeat **step 5** once more.
7. Remove tube from magnetic rack and resuspend in 11 μL per sample Bead Activation Buffer.

8. Transfer 10 μL ConA beads per sample into separate PCR tubes. Place on ice and used within 4 h of activation.
9. Add 100 μL nuclei suspension to ConA beads. Vortex at 800–1200 rpm for 5 s to evenly distribute beads. Incubate at room temperature for 10 min. Place tubes on magnetic rack for 2 min then, once clear, discard supernatant.
10. Remove tubes from magnetic rack. Resuspend in 50 μL Antibody Buffer by gently pipetting.
11. Examine bead binding as in Subheading 3.2.
12. Resuspend K-MetStat Panel by gently pipetting. Add 2 μL to positive and negative controls for downstream assessment of antibody specificity and sample quality (*see* **Note 3**). If using <100,000 nuclei, dilute the K-MetStat Panel in Antibody Buffer accordingly (e.g., 1:2 dilution for 50,000 nuclei).
13. Add 1 μL positive control antibody and 1 μL IgG negative control antibody to the respective samples. Add 0.1–0.5 μg primary antibody to all remaining samples.
14. Vortex at 800–1200 rpm for 5 s to evenly distribute beads. Incubate at 4 °C overnight on a nutator with the caps elevated (do not rotate tubes).

3.4 CUT&Tag Day 2 (Modified from CUTANA™ CUT&Tag Kit 2.0 User Manual)

1. Prepare Tagmentation Buffer for n samples plus positive and negative controls. Keep Wash Buffer 1, Wash Buffer 2 and Tagmentation Buffer on ice. Prepare 50 μL Pre-Wash Buffer and 500 μL 85% ethanol per sample. Keep at room temperature.
2. Remove sample tubes from 4 °C. Place tubes on magnetic rack for 2 min then, once clear, discard supernatant.
3. Remove tubes from magnetic rack and resuspend in 50 μL Wash Buffer 1 by gently pipetting.
4. Add 0.5 μL anti-rabbit IgG secondary antibody. Vortex at 800–1200 rpm for 5 s to evenly distribute beads. Incubate at room temperature for 30 min on a nutator with the caps elevated. Place tubes on magnetic rack for 2 min then, once clear, discard supernatant.
5. Keeping tubes on magnetic rack, wash twice with 200 μL Wash Buffer 1.
6. Remove tubes from magnetic rack and resuspend in 50 μL Wash Buffer 2 by gently pipetting.
7. Add 2.5 μL pAG-Tn5. Vortex at 800–1200 rpm for 5 s to evenly distribute beads. Incubate at room temperature for 1 h on a nutator with the caps elevated.
8. Meanwhile, pre-heat the thermocycler—block heated to 37 °C, lid heated to 47 °C.

9. Place tubes on magnetic rack for 2 min then, once clear, discard supernatant.
10. Remove tubes from magnetic rack and resuspend in 200 μL Wash Buffer 2 by gently pipetting. Place tubes on magnetic rack for 2 min then, once clear, discard supernatant. Repeat once more.
11. Remove tubes from magnetic rack and resuspend in 50 μL Tagmentation Buffer by gently pipetting.
12. Incubate in pre-heated thermocycler (block heated to 37 °C, lid heated to 47 °C) for up to 1 h (*see* **Note 4**).
13. Pre-heat another thermocycler—block heated to 58 °C, lid heated to 68 °C.
14. Remove tubes from 37 °C thermocycler. Place tubes on magnetic rack for 2 min then, once clear, discard supernatant.
15. Remove tubes from magnetic rack and resuspend in 50 μL room temperature Pre-Wash Buffer by very gently pipetting 3–5 times (beads may become clumpy after this step). Place tubes on magnetic rack for 2 min then, once clear, discard supernatant.
16. Remove tubes from magnetic rack. Add 5 μL SDS Release Buffer directly over the beads without disturbing with the pipette tip. Vortex at 2000 rpm for 5–10 s to resuspend.
17. Incubate in pre-heated thermocycler (block heated to 58 °C, lid heated to 68 °C) for 1 h.
18. Remove tubes from 58 °C thermocycler. Add 15 μL SDS Quench Buffer and vortex at 2000 rpm for 5–10 s to resuspend (do not pipette).
19. Add 2 μL of assigned i5 and i7 indexing primers to each sample, followed by 25 μL Non-Hot Start 2X PCR Master Mix. Vortex at 800–1200 rpm for 5–10 s to mix.
20. Perform PCR (lid heated to 105 °C) using the parameters in Table 1.
21. Remove tubes from thermocycler.
22. Vortex SPRIselect reagent at 2000 rpm until fully resuspended. Add 65 μL SPRIselect reagent and vortex at 800–1200 rpm for 5–10 s to mix. Repeat resuspension of SPRIselect reagent by vortexing between every four samples. Incubate at room temperature for 5 min. Place tubes on magnetic rack for 2 min then, once clear, discard supernatant.
23. Keep tubes on magnetic rack and wash twice with 200 μL 85% ethanol. Use a P10 pipette to remove any residual ethanol from the bottom of the tubes (do not quick spin).

Table 1
Thermocycling conditions for PCR

Step #	Temperature (°C)	Time	Number of cycles	Notes
1	58	5 min	1	Fill-in step
2	72	5 min	1	Extension
3	98	45 s	1	DNA melting
4 5	98 60	15 s 10 s	14–20 (*see* **Note 5**)	DNA melting Annealing/extension
6	72	1 min	1	Final extension
7	4	∞	1	Hold

24. Remove tubes from magnetic rack, leave caps open and allow to air-dry at room temperature for 2 min. Beads should appear dark brown and shiny (do not allow beads to become over-dry and cracked).
25. Add 17 μL 0.1X TE Buffer directly over the beads without disturbing with the pipette tip and vortex at 800–1200 rpm for 5–10 s to resuspend. Incubate at room temperature for 2 min. Place tubes on magnetic rack for 2 min until clear.
26. Transfer 15 μL supernatant to fresh PCR tubes. Proceed to library analysis or store libraries at −20 °C.

3.5 Library Analysis & Sequencing

1. Analyse library fragment size using the Agilent Bioanalyzer with the High Sensitivity DNA Kit according to the manufacturer's instructions. Bioanalyzer profiles should show a predominant enrichment for ~300–350 bp fragments (*see* **Note 4**). Lower yields are expected for experiments on root nuclei (*see* **Note 6**).
2. Quantify libraries using the Qubit fluorometer with the dsDNA HS Assay kit according to the manufacturer's instructions (if required).
3. Sequence libraries using an appropriate Illumina platform (e.g., NovaSeq 6000) to a depth of five to ten million paired-end reads per library.

3.6 Data Analysis

In this section, we outline a bioinformatic pipeline for processing and analysing paired-end sequencing data from raw fastq files, which will allow visualisation and comparison of average read distributions over genes of interest.

1. Download (and unzip) fastq files into working directory. Change.fastq extensions to.fq as required.
2. Trim adapters and remove low-quality reads using trim galore:

```
$ trim_galore --output_dir trimmedReads --length 30 --quality
20 --stringency 1 -e 0.1 --fastqc --paired {sampleName}_1.fq
{sampleName}_2.fq
```

3. Align trimmed reads to the reference genome using bowtie2 [6]:

```
$ bowtie2-build {genome}.fasta genomeIndex
$ bowtie2 --local --very-sensitive --no-mixed --no-discordant
--phred33 -I 10 -X 700 -x genomeIndex -1 trimmedReads/{sample-
Name}_1_val_1.fq -2 trimmedReads/{sampleName}_2_val_2.fq -S
{sampleName}.sam > {sampleName}_bowtie2.txt
```

Inspect the output.txt file containing a summary of the alignment. Three to five million uniquely mapped reads are ideal for downstream analysis.

4. Sort and remove PCR duplicates using the Picard Toolkit:

```
$picardCMD="java -jar picard.jar"
$ $picardCMD SortSam -I {sampleName}.sam -O {sampleName}.
sorted.sam -SORT_ORDER coordinate
$ picardCMD MarkDuplicates -I {sampleName}.newSorted.sam -O
{sampleName}.sorted.rmDup.sam -REMOVE_DUPLICATES true
```

5. Filter for high quality reads and index output BAM file using samtools [7]:

```
$ samtools view -hbS -F 0x04 -q 2 {sampleName}.sorted.rmDup.
sam > {sampleName}.bam
$ samtools index {sampleName}.bam
```

6. Convert to BIGWIG file and generate profile plots using deepTools [8] to analyse normalised read distribution over genes of interest (Fig. 1).

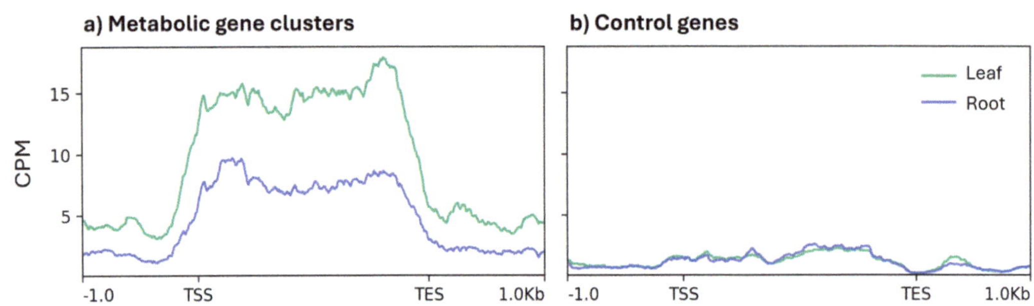

Fig. 1 H3K27me3 profile plot showing the average normalised read count (CPM) from the transcription start site (TSS) to the transcription end site (TES) plus 1 kb up- and down-stream for (**a**) root-expressed BGCs (no. of genes = 62) and (**b**) randomly selected control genes (no. of genes = 100) in root tissue (purple) versus leaf tissue (green)

```
$ bamCoverage --bam {sampleName}.bam -o {sampleName}.bw -of bigwig --normalizeUsing CPM
$ computeMatrix scale-regions -R {genesOfInterest}.bed -S {sampleName}.bw --regionBodyLength 2000 -a 1000 -b 1000 -bs 10 -o {sampleName}_matrix.gz
$ plotProfile -m {sampleName}_matrix.gz --outFileName {sampleName}_profilePlot.png
```

7. Sort and convert to BED format:

```
$ bedtools bamtobed -i {sampleName}.bam -bedpe > {sampleName}.bed
```

8. Identify significant peaks using preferred peak calling tool. We recommend csaw [9] for optimal peakcalling sensitivity, but others are available (*see* **Note** 7).
9. Visualise peakcalling output in the Integrated Genome Viewer (IGV; Fig. 2) [10] and perform statistical analyses, e.g., differential enrichment analysis can be performed using csaw.

4 Notes

1. To prevent damaging the nuclei, widen the bore of 20–200 μL pipette tips by trimming 2–3 mm from the end prior to use in resuspension steps.
2. Do not use ConA beads if they appear black, clumpy, or granular (may be caused by freezing or drying out). To prevent sample loss, use low-retention filter tips (e.g., S1120-8810, Starlab) when pipetting ConA beads.
3. To validate CUT&Tag experiments, positive and negative controls containing the K-MetStat Panel should be analysed using the SNAP-CUTANA_K-MetStat_Panel_Shell Script. Successful experiments show high antibody specificity and high signal-to-noise ratios, as seen in Fig. 3.
4. Tagmentation efficiency may vary for different primary antibodies. To enrich for ~300–350 bp fragments and avoid under- or over-tagmenting (Fig. 4), which may have negative consequences for downstream mapping and peakcalling, tagmentation time should be optimised for each primary antibody.
5. The number of PCR cycles should be optimised to obtain library concentrations sufficient for sequencing with the chosen Illumina platform while minimising PCR duplicates.

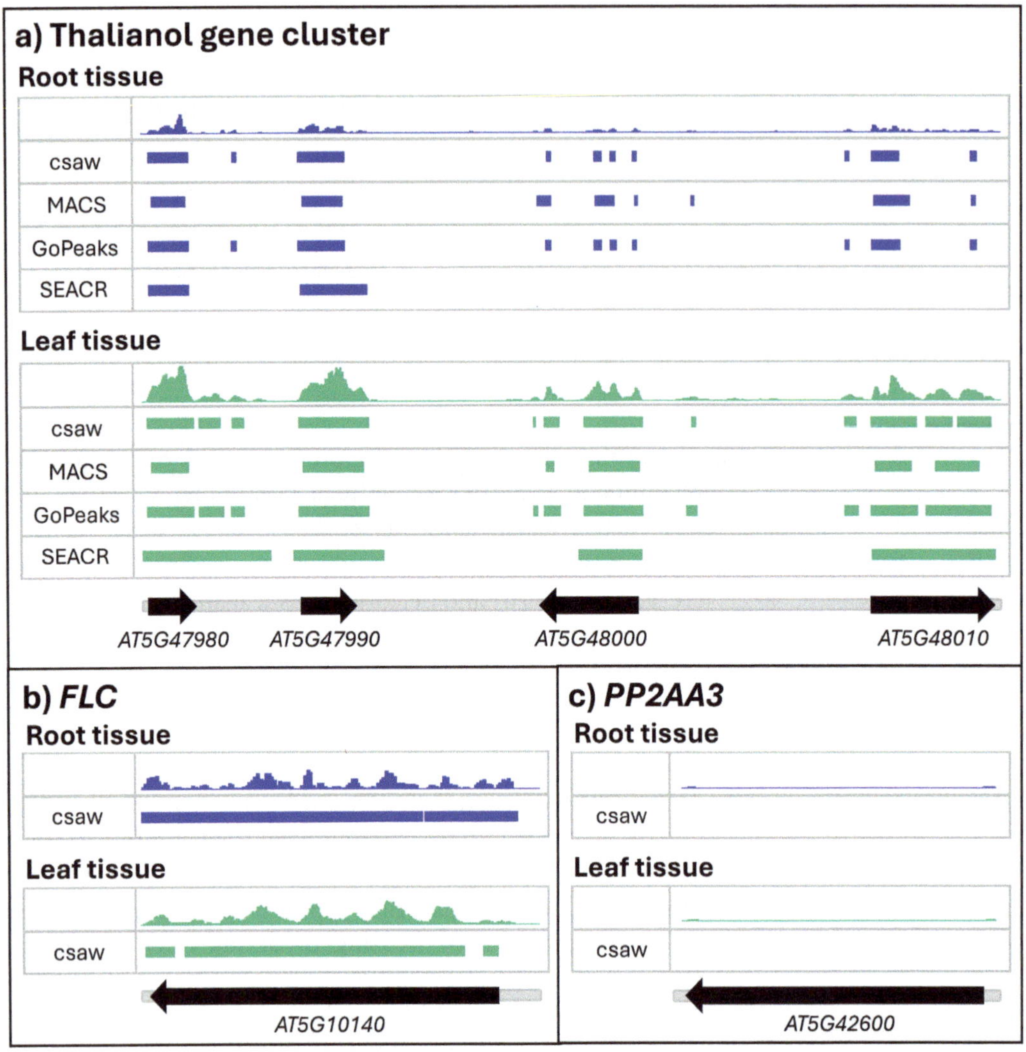

Fig. 2 IGV plots showing normalised read distribution in counts per million (CPM; upper track) and peakcalling outputs (lower tracks) from analysis of H3K27me3 CUT&Tag libraries in root (purple) and leaf (green). (**a**) Root-expressed thalianol cluster genes show greater H3K27me3 enrichment in leaves than roots. Csaw and GoPeaks exhibit greater sensitivity and resolution in peak detection relative to MACS and SEACR. (**b**) The constitutively repressed *FLC* gene shows similar H3K27me3 enrichment between root and leaf, serving as a positive control. (**c**) The constitutively active *PP2AA3* housekeeping gene shows similar depletion for H3K27me3 between the root and leaf, serving as a negative control

6. Lower yields are expected for experiments on root nuclei, but this does not significantly impact sequencing and downstream analyses (Fig. 5).

7. Alternative peakcalling tools are available, including macs2 [11], designed for ChIP-seq, SEACR [12], designed for CUT&RUN, and GoPeaks [13], designed for CUT&Tag. Evaluation of the outputs from each tool revealed that

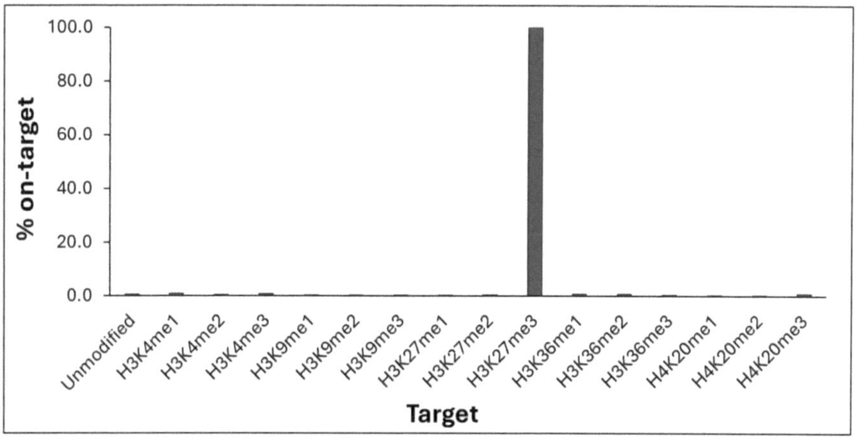

Fig. 3 Output of K-MetStat Panel analysis on exemplar H3K27me3 CUT&Tag library. The SNAP-CUTANA_K-MetStat_Panel_Shell Script aligns sequencing reads to unique DNA barcodes corresponding to each modified nucleosome in the panel (*x*-axis). Results inputted into the SNAP-CUTANA_K-MetStat_Panel_Analysis file are expressed as percent read recovery of each barcode relative to the read recovery for the target nucleosome (*y*-axis)

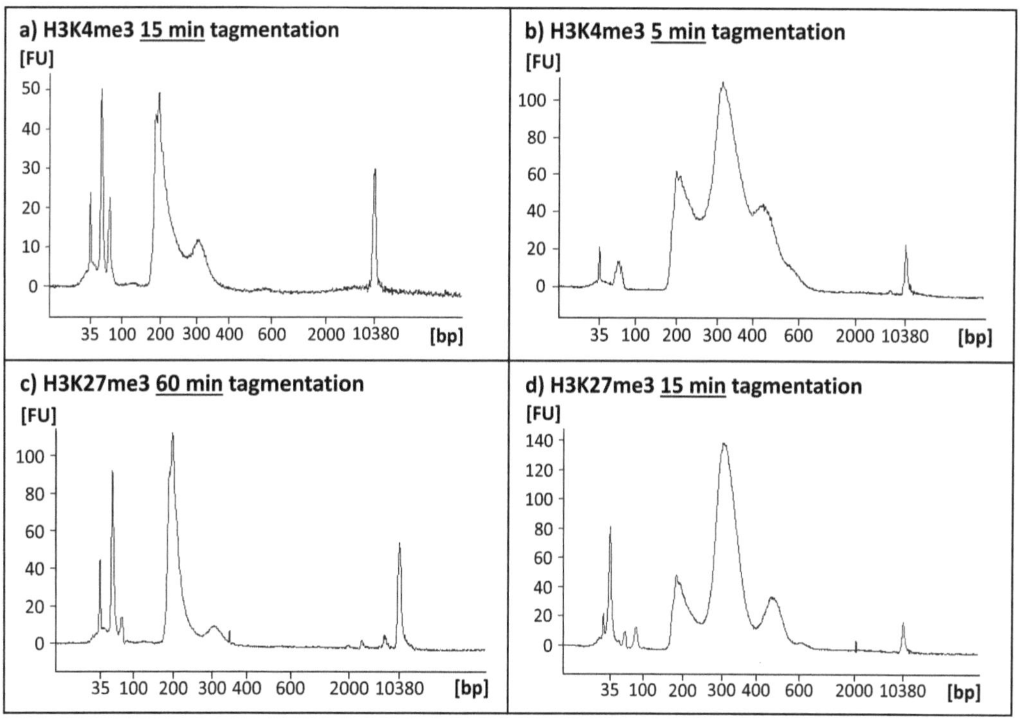

Fig. 4 Examples of "acceptable" (**a**, **c**) and "optimal" (**b**, **d**) bioanalyzer profiles for leaf CUT&Tag libraries using the anti-H3K27me3 (Cell Signaling Technology #9733) and anti-H3K4me3 (Abcam #8580) antibodies with indicated tagmentation times. Ideal bioanalyzer profiles show a predominant enrichment at 300–350 bp

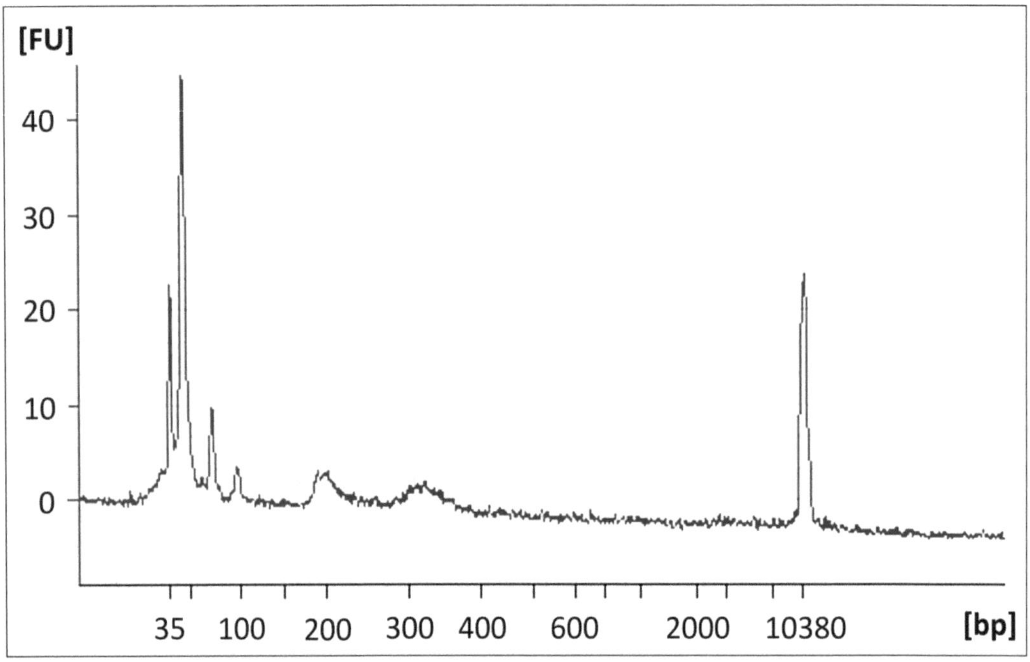

Fig. 5 Example bioanalyzer profile for a root CUT&Tag library using the anti-H3K27me3 antibody (Cell Signaling Technology #9733)

optimal sensitivity and resolution can be achieved using csaw or GoPeaks, which implement a window-based peakcalling approach. GoPeaks is a more user-friendly command line tool, while csaw requires more experience in R programming. However, csaw permits greater flexibility in parameter optimisation and a more rigorous statistical framework for differential enrichment analysis.

Acknowledgments

This work was supported by Royal Society awards RGF\R1\181018, RF\ERE\221026, and URF/R/221001 to H.W.N, the University of Exeter (H.W.N.), and the University of Bath (J.T. and H.W.N).

References

1. Nützmann HW, Huang A, Osbourn A (2016) Plant metabolic clusters – from genetics to genomics. New Phytol 211:771–789
2. Bharadwaj R, Kumar SR, Sharma A, Sathishkumar R (2021) Plant metabolic gene clusters: evolution, organization, and their applications in synthetic biology. Front Plant Sci 12: 697318
3. Yu N, Nützmann HW, Macdonald JT, Moore B, Field B, Berriri S et al (2016) Delineation of metabolic gene clusters in plant genomes by chromatin signatures. Nucleic Acids Res 44:2255–2265
4. Nutzmann HW, Doerr D, Ramirez-Colmenero A, Sotelo-Fonseca JE, Wegel E, Di Stefano M et al (2020) Active and repressed

biosynthetic gene clusters have spatially distinct chromosome states. Proc Natl Acad Sci USA 117:13800–13809

5. Liu Z, Cheema J, Vigouroux M, Hill L, Reed J, Paajanen P et al (2020) Formation and diversification of a paradigm biosynthetic gene cluster in plants. Nat Commun 11:5354
6. Langmead B, Salzberg SL (2012) Fast gapped-read alignment with Bowtie 2. Nat Methods 9: 357–359
7. Danecek P, Bonfield JK, Liddle J, Marshall J, Ohan V, Pollard MO et al (2021) Twelve years of SAMtools and BCFtools. GigaScience 10: giab008
8. Ramirez F, Ryan DP, Gruning B, Bhardwaj V, Kilpert F, Richter AS et al (2016) deepTools2: a next generation web server for deep-sequencing data analysis. Nucleic Acids Res 44:W160–W165
9. Lun ATL, Smyth GK (2016) csaw: a Bioconductor package for differential binding analysis of ChIP-seq data using sliding windows. Nucleic Acids Res 44:5
10. Thorvaldsdottir H, Robinson JT, Mesirov JP (2013) Integrative Genomics Viewer (IGV): high-performance genomics data visualization and exploration. Brief Bioinform 14:178–192
11. Zhang Y, Liu T, Meyer CA, Eeckhoute J, Johnson DS, Bernstein BE et al (2008) Model-based analysis of ChIP-Seq (MACS). Genome Biol 9:R137
12. Meers MP, Tenenbaum D, Henikoff S (2019) Peak calling by Sparse Enrichment Analysis for CUT&RUN chromatin profiling. Epigenetics Chromatin 12:42
13. Yashar WM, Kong G, VanCampen J, Curtiss BM, Coleman DJ, Carbone L et al (2022) GoPeaks: histone modification peak calling for CUT&Tag. Genome Biol 23:144

Chapter 13

Mapping the 3D Chromosome Organization of *Streptomyces* by Hi-C

Zhiyu Zhong, Lin Liu, Liang Deng, Zhihu Zhao, and Yuhui Sun

Abstract

Streptomyces species are prolific producers of bioactive compounds, including antibiotics, but many of their biosynthetic gene clusters (BGCs) remain silent or are poorly expressed under laboratory conditions. Recent studies have reported a close correlation between chromosomal organization and the expression activity of genes, including BGCs. Chromosome conformation capture with high-throughput sequencing (Hi-C) is a technique that allows for the profiling of three-dimensional genome organization by capturing physical interactions between chromosomal regions. In this chapter, we describe the application of Hi-C to *Streptomyces coelicolor*, focusing on the detailed protocol for constructing a Hi-C library and analyzing chromosomal interactions. By using this technique, researchers can explore the relationship between chromosomal architecture and the expression of BGCs, which may inform new strategies for improving natural product yields in *Streptomyces*.

Key words *Streptomyces*, Chromosomal organization, Hi-C, Library preparation, Data analysis

1 Introduction

Streptomyces are widely recognized for their ability to produce an array of bioactive compounds, many of which play crucial roles in medicine and agriculture. These organisms are especially notable for harboring biosynthetic gene clusters (BGCs) responsible for producing antibiotics and other secondary metabolites [1, 2]. However, a significant challenge remains: many BGCs are either silent or poorly expressed under standard laboratory conditions, limiting their potential for natural product discovery.

The rise of antibiotic resistance has made the discovery of novel antibiotics and optimization of existing ones increasingly urgent [3]. While several strategies, including heterologous expression of BGCs and manipulation of transcriptional regulators, have been explored [4], the influence of chromosomal organization on gene expression is often overlooked. It has been shown that the linear

Hans-Wilhelm Nützmann and Ramesha Thimmappa (eds.), *Metabolic Gene Clusters: Methods and Protocols*, Methods in Molecular Biology, vol. 3054, https://doi.org/10.1007/978-1-0716-5384-5_13,

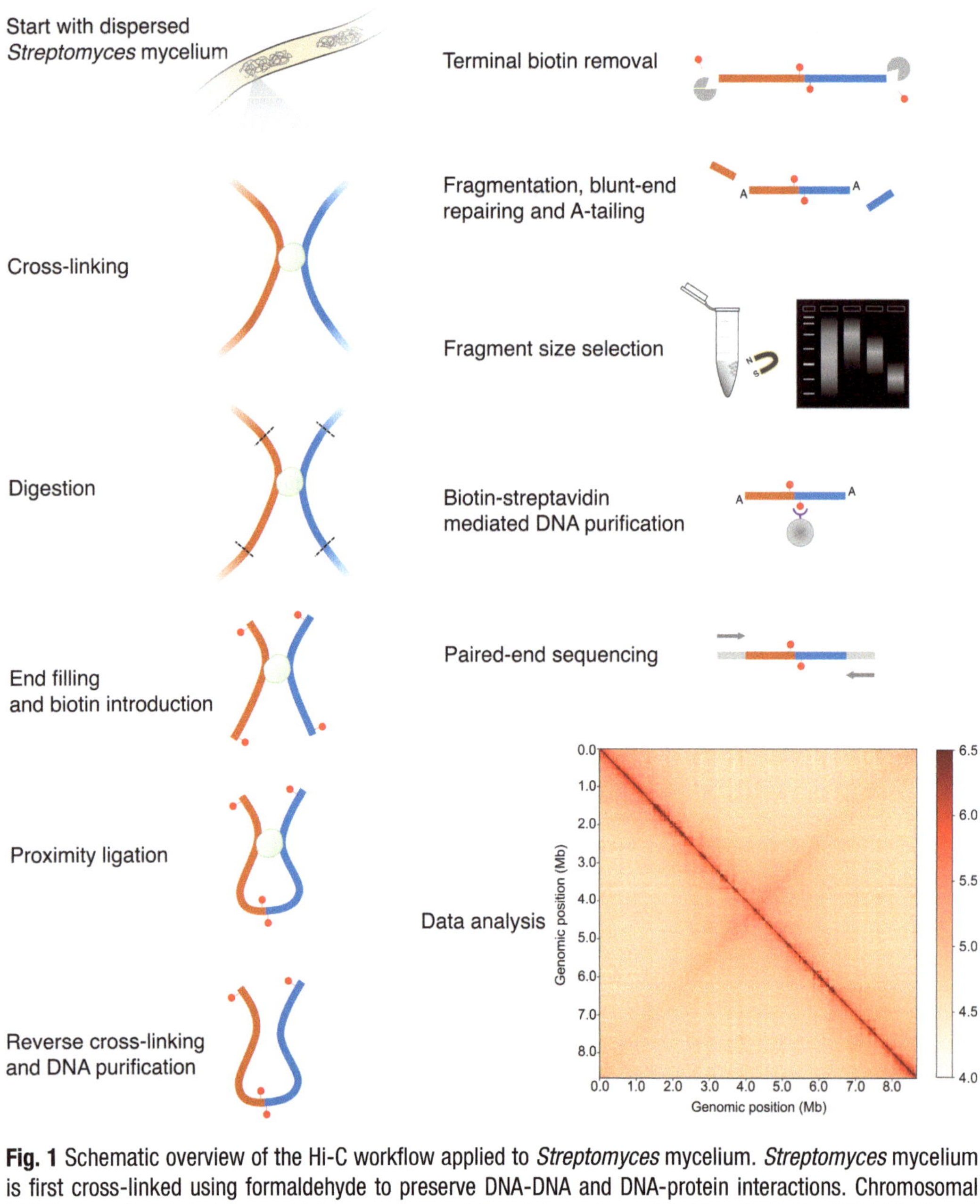

Fig. 1 Schematic overview of the Hi-C workflow applied to *Streptomyces* mycelium. *Streptomyces* mycelium is first cross-linked using formaldehyde to preserve DNA-DNA and DNA-protein interactions. Chromosomal DNA is then digested with a restriction enzyme, and the sticky ends are filled in with biotin-dCTP. Ligation of DNA fragments that were spatially close in the cell creates ligated products representing genomic interactions. After reversing the cross-links, the DNA is purified, and unligated biotin ends are removed. The DNA is then fragmented, repaired, A-tailed, and size-selected. Biotin-labeled fragments are enriched using streptavidin-coated magnetic beads and subjected to paired-end sequencing. The sequencing data are analyzed to generate a contact map, revealing the spatial organization of the genome

position of genes within the chromosome can significantly influence their expression levels, a phenomenon known as the chromosomal position effect [5, 6], but the underlying mechanisms are not fully understood.

Chromosome conformation capture (3C) related techniques including Hi-C (Fig. 1) reveal how often different chromosomal regions physically interact, enabling researchers to map the genome's spatial organization in detail [7–9]. Recent applications of Hi-C to *Streptomyces* have advanced our understanding of these bacteria. In *Streptomyces venezuelae*, Hi-C studies have shown that the spatial rearrangement of its linear chromosome occurs during sporulation, where chromosome arms are brought into proximity by segregation proteins, and terminal domains are organized by nucleoid-associated proteins, facilitating proper chromosome compaction and segregation [10]. Similarly, a study on *Streptomyces ambofaciens* revealed that the chromosome undergoes structural rearrangement during metabolic differentiation. This shift from an "open" to "closed" conformation corresponds with the activation of BGCs, indicating a correlation between chromosome structure and the biosynthesis of specialized metabolites [11]. In our recent work, we applied high-resolution Hi-C to *Streptomyces coelicolor* to investigate chromosomal reorganization during the transition from primary to secondary metabolism. We discovered that the chromosome undergoes global structural changes, correlating with BGC activation. Using the frequently interacting regions (FIREs) metric, we demonstrated that gene expression was strongly linked to local interaction frequencies. By integrating an exogenous reporter gene and an RK-682 BGC into chromosomal loci with high FIRE values, we successfully enhanced gene expression and natural product yields [12]. These studies highlight the interplay of three-dimensional chromosome organization and BGC expression and present a new strategy for improving natural product biosynthesis in *Streptomyces* through 3D structure-guided targeted genomic manipulation. It is also noteworthy that a recent study on fungi revealed that active BGCs form jet-like chromatin structures [13], suggesting that changes in chromatin organization may play a broader role in regulating secondary metabolism across species.

In this chapter, we provide a step-by-step protocol for Hi-C on *S. coelicolor*, covering bacterial culture, Hi-C library construction, and chromosomal interaction analysis. The resolution of the Hi-C results largely depends on the choice of restriction enzyme. Here, we use *Sau*3AI, a 4-cutter that generates high-resolution maps by cutting DNA frequently, though it requires deep sequencing to ensure sufficient coverage. For capturing larger chromatin structures or lower-resolution analysis, other enzymes can be used. While optimized for *S. coelicolor*, this method is flexible and can be adapted to other *Streptomyces* strains. We also recommend referring to references [10, 11] and an earlier bacterial Hi-C protocol [14] for foundational understanding and guidance for modifications.

2 Materials

1. *Streptomyces* strains of interest.
2. Soy flour-mannitol (SFM) agar medium: 33 g/L soya flour, 20 g/L mannitol, 20 g/L agar. Autoclave twice at 115 °C for 15 min each time, with gentle shaking between the two rounds.
3. 2 × TY medium: 16 g/L tryptone, 10 g/L yeast extract, 5 g/L NaCl. Autoclave at 121 °C for 15 min (*see* **Note 1**).
4. Trace element solution: 40 mg/L $ZnCl_2$, 200 mg/L $FeCl_3{\cdot}6H_2O$, 10 mg/L $CuCl_2{\cdot}2H_2O$, 10 mg/L $MnCl_2{\cdot}4H_2O$, 10 mg/L $Na_2B_4O_7{\cdot}10H_2O$ and 10 mg/L $(NH_4)_6Mo_7O_{24}{\cdot}4H_2O$. Filter-sterilize, store at 4 °C.
5. R5- medium: 5.73 g/L TES, 103 g/L sucrose, 10 g/L glucose, 5 g/L yeast extract, 10.12 g/L $MgCl_2{\cdot}6H_2O$, 0.25 g/L K_2SO_4, 0.1 g/L casamino acids, 2 mL/L trace element solution. Adjust pH to 6.8 with 5 M NaOH. Autoclave at 115 °C for 20 min (*see* **Note 1**).
6. 20% (v/v) Glycerol: Autoclave at 121 °C for 15 min (*see* **Note 2**).
7. Antifoam 204 (Sigma).
8. 37% Formaldehyde.
9. 80% Ethanol.
10. 2 M Glycine: Dissolve 7.5 g glycine in 50 mL ddH_2O, store at 4 °C.
11. SET buffer: 75 mM NaCl, 25 mM EDTA pH 8, 20 mM Tris-HCl, pH 7.5.
12. TE buffer: 10 mM Tris-HCl, pH 8, 1 mM EDTA pH 8.
13. 60 mg/mL Lysozyme.
14. 10% (w/v) sodium dodecyl sulfate (SDS).
15. 10 × NEBuffer r2.1(NEB): 500 mM NaCl, 100 mM Tris-HCl, 100 mM $MgCl_2$, 1 mg/mL recombinant albumin, pH 7.9.
16. Triton neutralization buffer: 6% (v/v) Triton X-100, 1 × NEBuffer r2.1.
17. Digestion buffer: 1% (v/v) Triton X-100, 1 × NEBuffer r2.1.
18. 5 units/μL *Sau*3AI restriction enzyme.
19. End-filling buffer: for 120 μL buffer, add 12 μL 10 × NEBuffer r2.1, 1.8 μL 10 mM dATP, 1.8 μL 10 mM dGTP, 1.8 μL 10 mM dTTP, 3.6 μL 5 mM biotin-14-dCTP, 3 μL 5 units/μL DNA polymerase I large (klenow) fragment and 96 μL ddH_2O.

20. Ligation buffer: for 880 μL buffer, add 50 μL 20% (v/v) Triton X-100, 100 μL 10 × NEBuffer r2.1, 5 μL 20 mg/mL BSA, 10 μL 5 units/μL T4 DNA ligase, and 715 μL ddH_2O.
21. QIAamp DNA Mini Kit (Qiagen) that includes Proteinase K, Buffer AL, Buffer AW1, Buffer AW2, Buffer AE and QIAamp Mini spin columns.
22. 3 units/μL T4 DNA polymerase.
23. VAHTS DNA Clean Beads (Vazyme).
24. VAHTS Universal Plus DNA Library Prep Kit for Illumina (Vazyme) that includes FEA Enzyme Mix, FEA Buffer, Rapid DNA Ligase, Rapid Ligation Buffer 3, VAHTS HiFi Amplification Mix, PCR Primer Mix, Neutralization Buffer, and Control DNA.
25. Dynabeads MyOne Streptavidin C1 (Invitrogen).
26. Equalbit 1 × dsDNA HS Working Solution (Vazyme).
27. Tween Wash Buffer (TWB): 5 mM Tris-HCl pH 8.0, 0.5 mM EDTA, 1 M NaCl, 0.05% Tween.
28. 2 × Binding Buffer (2 × BB): 10 mM Tris-HCl pH 8.0, 1 mM EDTA, 2 M NaCl.
29. VAHTS DNA Adapter for Illumina (Vazyme).
30. Sterilized cotton swab.
31. Centrifuge tubes.
32. Incubator.
33. Shaking incubator.
34. Water bath.
35. Microscope.
36. Rotating mixer.
37. Magnetic separation device.
38. Qubit fluorometer.
39. Centrifuge.
40. Vortex.
41. Thermal cycler.

3 Methods

3.1 Streptomyces *Culture*

1. Prepare SFM agar plates by pouring 20 mL of SFM agar medium into each petri dish. Air dry the plates with the lids open in a laminar flow hood for 40 min to ensure the surface is dry, accommodating the hydrophobic nature of *Streptomyces* mycelium.

2. Streak *Streptomyces* onto the SFM agar plate using a sterile loop or flat wooden stick. Use a pattern that thins out the sample and separates individual cells. Incubate the plates at 28 °C for 7 days to allow the formation of single colonies. Monitor for different colony morphologies and contamination by other bacteria or fungi.
3. Inoculate a single clone onto a fresh SFM agar plate for expansion, and incubate at 28 °C for 10–14 days to allow for sporulation (*see* **Note 3**).
4. Harvest the spores gently using a wet cotton swab, and release the spores by rotating the swab inside a 15 mL centrifuge tube containing 10 mL of water while pressing it firmly against the inner wall.
5. Evaluate spore density by hemocytometer counting method, adjusting the dilution gradient as needed (up to a 1:100 dilution).
6. Pellet the spores by centrifuge at 1000× *g* for 5 min, and discard the supernatant.
7. Resuspend the spores in 20% glycerol and store frozen at −20 °C (*see* **Note 4**).
8. When thawing, take the cryovial out of the storage refrigerator and place it in a 37 °C water bath for 3 min, making sure the cap remains above the water level (*see* **Note 5**).
9. Transfer the spores into a 1.5 mL centrifuge tube and centrifuge at 1000× *g* for 5 min to pellet the spores and discard the supernatant.
10. Wash the spores with 1 mL of water. Centrifuge again at 1000× *g* for 5 min.
11. Inoculate 5×10^9 spores into 50 mL of 2 × TY medium in a 250 mL flask containing a coiled spring. Incubate at 30 °C on a shaker at 220 rpm for 8 h, until emerging germ tubes are visible under microscope (*see* **Note 6**).
12. Pellet the germinated spores by centrifugation at 1000× *g* for 5 min and discard the supernatant.
13. Transfer 5×10^9 germinated spores into 500 mL of R5-medium in a baffled 2 L flask to make an inoculum of 1×10^7 spores/mL, the desired final OD_{450} is approximately 0.06 against the R5-medium (volumes can be adjusted based on the culture system).
14. Add one drop of sterilized antifoam 204, then incubate at 30 °C on a shaker at 220 rpm until the desired time point for detection.

3.2 Hi-C Library Construction

3.2.1 Cross-Linking

1. Separate the cells from the supernatant by centrifuging at 1000× *g* for 5 min at room temperature (RT).
2. Prepare a 4 mL suspension of the mycelium using the supernatant, adjust the OD_{450} to approximately 0.2.
3. Add 37% formaldehyde to achieve a final concentration of 3%, and cross-link at 4 °C with gentle agitation for 1 h.
4. Quench the reaction by adding 2 M glycine to achieve a final concentration of 375 mM, and incubate for 15 min at 4 °C with gentle agitation.
5. Pellet the fixed mycelium by centrifuging at 1000× *g* for 5 min at 4 °C, and discard the supernatant.
6. Wash the mycelium with 5 mL of SET buffer, centrifuge at 1000× *g* for 5 min at 4 °C, and discard the supernatant.
7. Resuspend the mycelium in 1 mL of SET buffer, and divide into two 500 μL aliquots (*see* **Note 7**).
8. Pellet the mycelium by centrifuging at 1000× *g* for 5 min at 4 °C, and discard the supernatant.
9. Quick-freeze the cell pellets in liquid nitrogen. Store the pellets at −80 °C or use immediately for further experiments.

3.2.2 Lysis and Digestion

1. Resuspend the mycelium in 500 μL of TE buffer.
2. Add 20 μL 60 mg/mL lysozyme, and incubate at 37 °C for 10–30 min to lyse the cells.
3. Add 27 μL of 10% SDS to inactivate the excess lysozyme, mix well, and incubate at 65 °C for 10 min.
4. Add 500 μL of Triton neutralization buffer to quench the additional SDS, and shake gently for 15 min at 37 °C.
5. Centrifuge at 2000× *g* for 5 min at 4 °C, and discard the supernatant (*see* **Note 8**).
6. Add 500 μL of digestion buffer and 50 units of *Sau*3AI, mix gently, and incubate at 37 °C for 1 h.
7. Take a 50 μL aliquot for quality control, checking the digestion efficiency by gel electrophoresis. Use cross-linked but undigested gDNA (Fig. 2, Lane 3) and uncross-linked but digested gDNA (Fig. 2, Lane 4) as controls, a smear pattern indicates effective digestion of cross-linked gDNA (Fig. 2, Lane 5).
8. Centrifuge at 2000× *g* for 5 min at 4 °C, and discard the supernatant.

3.2.3 End-Filling and Ligation

1. Add 120 μL of end-filling buffer to the digested products, mix well, and incubate with gentle shaking at 37 °C for 55 min.
2. Inactivate the Klenow fragment by heating at 65 °C for 20 min. Immediately place the sample on ice to cool.

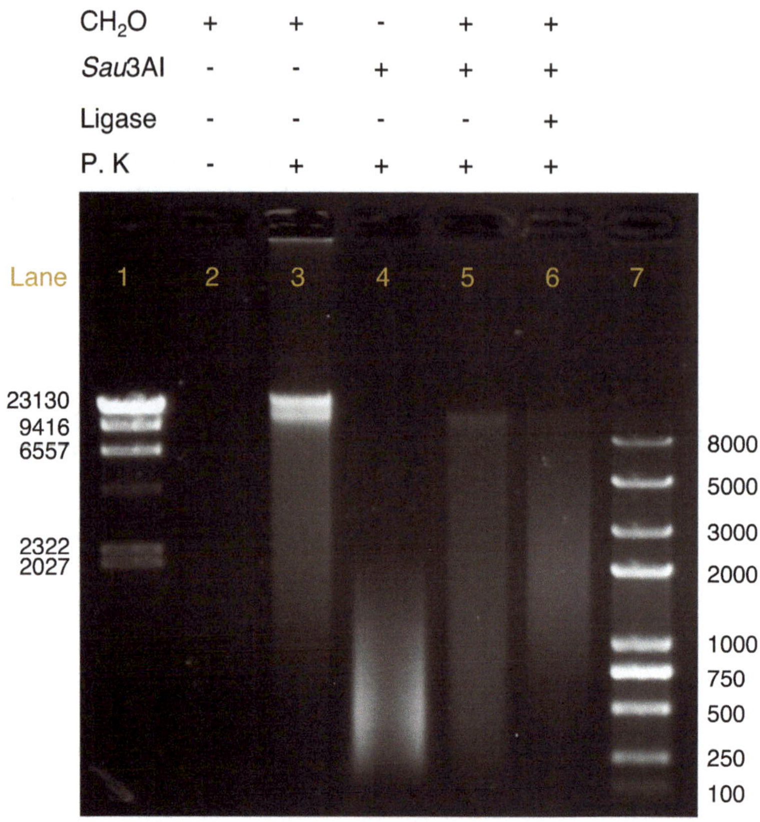

Fig. 2 Quality control of Hi-C library preparation. Lane 1 and 7: DNA Ladder. Lane 2: Cross-linked genomic DNA without Proteinase K treatment (without reverse cross-linking). Lane 3: Cross-linked genomic DNA with Proteinase K treatment. Lane 4: Uncross-linked genomic DNA digested with *Sau*3AI. Lane 5: Cross-linked genomic DNA digested with *Sau*3AI. Lane 6: Ligation product

3. Add 880 μL of ligation buffer, mix gently, and incubate at 16 °C for 4 h. Invert the tubes gently every 1 h to ensure thorough mixing. Take a 50 μL aliquot for quality control, checking the ligation efficiency by gel electrophoresis. An upward shift of the main DNA smear indicates an increase in fragment size, confirming successful ligation (Fig. 2, Lane 6).
4. Centrifuge the sample at 2000× *g* for 5 min at 4 °C, and discard the supernatant.

3.2.4 Reverse Cross-Linking and DNA Purification

1. Reverse cross-linking and DNA purification can be performed using the QIAamp DNA Mini Kit. Add 180 μL Buffer AL and 20 μL of 2 mg/mL proteinase K, incubate the sample at 55 °C for 1–3 h to degrade residual proteins.
2. After column purification, elute DNA with 100 μL Buffer AE (*see* **Note 9**).

3.2.5 Terminal Biotin Removal

1. The unligated terminal biotin could be removed using T4 DNA polymerase. Pipette 86.7 μL purified ligation product into a new PCR tube. Add 10 μL 10 × NEBuffer r2.1, 1 μL 10 mM dATP, 1 μL 10 mM dTTP, 1.3 μL T4 DNA polymerase, and then mix and incubate at 12 °C for 4 h.
2. Add 100 μL VAHTS DNA Clean Beads. Mix and incubate for 5 min at RT.
3. Centrifuge briefly, then place the PCR tube on magnetic separation device, discard the supernatant after the solution is clarified (about 3 min).
4. Keep the PCR tube on magnetic separation device. Add 200 μL newly prepared 80% ethanol to wash the beads. Incubate for 30 s at RT. Discard the supernatant.
5. Repeat **step 4** once.
6. Keep the PCR tube on magnetic separation device, discard the residual ethanol at the bottom of the tube. Air dry until the beads are not reflective.
7. Add 36 μL ddH_2O, mix by pipetting up and down. Incubate for 5 min at RT off the magnet separation device.
8. Centrifuge briefly. Then place the PCR tube on magnetic separation device. Transfer 35 μL solution into a new PCR tube.
9. Measure the DNA concentration by Qubit fluorometer (*see* **Note 10**).

3.2.6 Fragmentation, Blunt-End Repairing and A-Tailing

1. Many vendors offer kits that can perform enzymatic DNA fragmentation, end repair, 5′-phosphorylation, and 3′-dA tailing in a one-pot reaction. In this protocol, we use FEA Enzyme Mix and FEA Buffer from the VAHTS Universal Plus DNA Library Prep Kit for Illumina. Prepare the reaction in a PCR tube on ice: 0.5–1 μg of the DNA sample from the previous section, 5 μL of FEA Buffer, and add ddH_2O to bring the final volume to 40 μL.
2. Add 10 μL of the FEA Enzyme Mix to the reaction, mix well, briefly centrifuge, and return the tube to ice.
3. Place the tube in a PCR instrument and run the following program: Hot lid at 105 °C, 37 °C for 30 min, 65 °C for 30 min, and hold at 4 °C (*see* **Note 11**).

3.2.7 Fragment Size Selection

1. DNA fragment size selection is performed using a two-step magnetic bead method. Vortex the VAHTS DNA Clean Beads to ensure thorough mixing once they have equilibrated to RT.

2. Add 50 μL of VAHTS DNA Clean Beads to the 50 μL A-tailed product. Mix well and incubate for 5 min at RT.
3. Place the PCR tube on a magnetic separation device, discard the supernatant after the solution is clarified (about 3 min).
4. Keeping the tube on the magnetic separation device, wash the beads with 200 μL of freshly prepared 80% ethanol. Incubate for 30 s at RT, then discard the supernatant.
5. Repeat the ethanol wash twice more.
6. Discard any residual ethanol at the bottom of the tube using a 10 μL pipette. Air dry the beads until they are no longer reflective.
7. Add 101 μL of ddH_2O and mix by pipetting up and down. Incubate for 5 min at RT off the magnet separation device.
8. Centrifuge briefly, place the tube back on the magnetic separation device, and transfer 100 μL of the supernatant to a new PCR tube.
9. Add 60 μL of VAHTS DNA Clean Beads to the supernatant, mix well, and incubate for 5 min at RT.
10. After brief centrifugation, place the tube on the magnetic separation device, transfer the supernatant to a new PCR tube, and discard the beads.
11. Add 12 μL of VAHTS DNA Clean Beads to the supernatant, mix well, and incubate for 5 min at RT.
12. Centrifuge briefly, place the tube on the magnetic separation device, and discard the supernatant while keeping the beads.
13. Wash the beads with 200 μL of freshly prepared 80% ethanol, incubate for 30 s at RT, and discard the supernatant.
14. Repeat the ethanol wash twice more.
15. Remove residual ethanol using a 10 μL pipette. Air dry the beads until no liquid remains visible.
16. Add 51 μL of ddH_2O, mix by pipetting up and down, and incubate for 5 min at RT off the magnet separation device.
17. Centrifuge briefly, place the tube back on the magnetic separation device, allow the solution to clarify (about 3 min), then transfer 50 μL of the supernatant to a new PCR tube.

3.2.8 Biotin-Streptavidin-Mediated DNA Purification

1. Vortex the Dynabeads MyOne Streptavidin C1 to ensure proper mixing. Transfer 5 μL of beads per sample to a new 1.5 mL tube.
2. Add 100 μL of TWB, mix thoroughly by pipetting up and down, and incubate for 3 min at RT on rotating mixer.

3. Centrifuge briefly, place the tube on a magnetic separation device, and discard the supernatant after the solution is clarified (about 3 min).
4. Repeat **steps 2** and **3** once more.
5. Resuspend the Streptavidin C1 beads in 50 μL of 2 × BB, then add the resuspended beads to 50 μL of the fragment size selection product. Incubate for 15 min at RT on a rotating mixer.
6. Centrifuge briefly, place the tube back on the magnetic separation device, and discard the supernatant after the solution is clarified (about 3 min).
7. Add 65 μL of ddH_2O, mix by pipetting up and down to resuspend the DNA and beads (*see* **Note 12**).

3.2.9 Adapter Ligation

1. Prepare the reaction in a PCR tube on ice: 65 μL Streptavidin C1 beads binding DNA, 25 μL Rapid Ligation Buffer 3, 5 μL Rapid DNA Ligase, and 5 μL VAHTS DNA Adapter for Illumina (*see* **Note 13**). Mix thoroughly and incubate the reaction at 20 °C for 15 min.
2. Centrifuge briefly, then place the PCR tube on a magnetic separation device, discard the supernatant after the solution is clarified (about 3 min).
3. Resuspend the beads in 100 μL of TWB, then transfer the suspension to a new PCR tube.
4. Centrifuge briefly, then place the new tube on the magnetic separation device, discard the supernatant after the solution is clarified (about 3 min).
5. Repeat **steps 3** and **4** twice more.
6. Add 11 μL of ddH_2O to resuspend the beads, and incubate at 70 °C for 5 min.
7. After incubation, centrifuge briefly and place the PCR tube on the magnetic separation device. Transfer 10 μL of the clarified supernatant to a new PCR tube.
8. Repeat **steps 6** and 7, get a total elution volume of 20 μL DNA.

3.2.10 PCR Amplification and Sequencing

1. Prepare the reaction in a PCR tube: for 20 μL of the adapter-ligated product, add 25 μL of VAHTS HiFi Amplification Mix and 5 μL of PCR Primer Mix.
2. Place the mix to thermal cycler, and run the following program: denaturation at 95 °C for 3 min, (denaturation at 98 °C for 20 s, primer annealing at 60 °C for 15 s, extension at 72 °C for 30 s) × 9 cycles (*see* **Note 14**), final extension at 72 °C for 5 min, hold at 4 °C.

3. Vortex the VAHTS DNA Clean Beads to ensure they are mixed after equilibration to RT.
4. Add 50 μL of beads to the amplification product. Mix and incubate for 5 min at RT.
5. Place the tube on a magnetic separation device, discard the supernatant after the solution is clarified (about 3 min).
6. Keeping the PCR tube on the magnetic separation device, wash the beads with 200 μL of freshly prepared 80% ethanol. Incubate for 30 s at RT, then discard the supernatant.
7. Repeat the ethanol wash two more times.
8. Remove residual ethanol using a 10 μL pipette, then air-dry the beads until they are no longer reflective.
9. Add 51 μL of ddH_2O, mix by pipetting up and down, and incubate for 5 min at RT.
10. Centrifuge briefly, then place the tube back on the magnetic separation device. Transfer 50 μL of the supernatant to a new PCR tube.
11. Repeat the purification **steps 4–8**.
12. Add 31 μL of ddH_2O, mix by pipetting up and down, and incubate for 5 min at RT.
13. Centrifuge briefly, place the tube back on the magnetic separation device, and transfer 30 μL of the clarified supernatant to a new 1.5 mL centrifuge tube.
14. Measure the DNA concentration of the Hi-C library by using the Equalbit 1 × dsDNA HS Working Solution and a Qubit fluorometer, and prepare the library for next-generation sequencing. Paired-end sequencing with read lengths of 36–150 bp can be performed on the Illumina or MGI sequencing platform. For prokaryotic samples, 5000×–10,000× coverage with paired-end 150 bp reads per replicate is sufficient.

3.3 Data Analysis

3.3.1 Contact Map Construction

The Hi-C contact map provides a framework for analyzing the spatial structure of the genome. To generate a contact map from raw Hi-C sequence data, the data processing processes include reads quality filtering and trimming, reference genome mapping, unusable data filtering, raw contact map construction, and contact map balancing.

Raw NGS data (either from Illumina or MGI) always contain low-quality reads and adapter sequences. For filtering, we recommend using Trimmomatic [15] with the parameter "ILLUMINACLIP" to specify adapter information and the parameter "MINLEN:36" to retain reads longer than 36 bp after trimming.

There are many standard software tools and pipelines available for Hi-C analysis, such as HiCUP [16], HiC-Pro [17], Chromap

[18], and Distiller (https://github.com/open2c/distiller-nf). We recommend the following: In the reads mapping step, use BWA [19] with the "MEM" function and the "-SP5M" parameter to align the reads, as BWA can directly map Hi-C chimeric reads to the reference genome and get all split alignment sites. In the data filtering step, use Pairtools (https://github.com/open2c/pairtools) to perform data filtering and deduplication, as Pairtools can extract the correct paired contact signals from the BWA mapping results. In the contact map construction and balancing steps, use Cooler [20] with bin sizes of 40 kb, 5 kb, and 2 kb.

3.3.2 FIRE Analysis

The FIRE value shows the local interaction frequency of the genome, based on Schmitt's method [21]. First, use the F_GC_MAP.file.sh script from the HiCnv software (https://github.com/ay-lab/HiCnv) to calculate the genome's effective fragment length, GC content, and mappability. Then, count the local interactions (sum of contacts in the distance range of 15 ~ 200 kb) for each bin from the 5 kb raw matrix. Second, use the HiCNormCis algorithm to calculate the FIRE value. HiCNormCis is a modified version of the HiCNorm code and can be found in the SI dataset 5 of our previously published work [12].

3.3.3 Comparison of FIRE Value with RNA-seq

First, z-transform and Gaussian smooth the FIRE-like signals (FIRE value) and RNA-seq signals (RNA-seq depth for each 5 kb bin) using Virginia's method [22]. For global comparison, use the total FIRE value and RNA-seq signal to perform Pearson correlation analysis. For local comparison, use the FIRE value and RNA-seq signals from each sliding window (11 continuous 5 kb bins) to perform Pearson correlation analysis.

3.3.4 Chromosomal Interaction Domains (CID) Analysis

CIDs can be calculated using standard software such as hicFindTADs from HiCExplorer [23], or by using Le's method [24]. Le's method is based on directional index (DI). To calculate DI, for each 5 kb bin, interactions between the bin and upstream/downstream bins within 200 kb (window size) are extracted to form two vectors. The DI is the *T*-value from a paired test of Log2 values of the two vectors. The CID boundary is defined where the bin's upstream DI value is below −1.81, and the downstream DI value is above 1.81 (*see* **Note 15**).

3.3.5 3D Chromosomal Structure Modeling

A 3D chromosomal structure model can be constructed using Pastis (v0.1) [25] or miniMDS [26] software with a multidimensional scaling (MDS) model. For Pastis software, briefly, a Numpy-format 5 kb balanced contact map is used as input. The pastis-mds script is then used to calculate the 3D model. The software outputs a file in.pdb (Protein Data Bank) format, where every 20 consecutive "O" atoms represent a 5 kb genomic sequence. Software tools for visualizing 3D biological macromolecules, such as PyMol [27], can be used for genome structure visualization.

3.3.6 A/B Compartment Analysis

A/B compartments can be calculated using standard software such as hicPCA from HiCExplorer [23] and CscoreTool (v1.1) [28]. A resolution of 40 kb can be selected. Gene density or RNA-seq signals can be used to refine the A/B compartment (*see* **Note 16**). In previous research, the compartment signal PC1 (eigenvector 1) correlates with the distribution of genes and features of open chromatin [9]. The A compartment indicates the active chromosomal region. Thus, gene density, RNA-seq, and active histone markers could be used to refine the A/B compartment. Briefly, gene density or average RNA-seq coverage of each 40 kb genome bin should first be calculated as an eigenvalue. Then, for each chromosome, compare the average eigenvalues of bins with positive compartment signals and negative signals. If the eigenvalue of bins with a positive compartment signal is less than that of bins with a negative compartment signal, the compartment signal of all the bins in the chromosome should be multiplied by −1. Finally, bins with a positive compartment signal represent the A compartment, and those with a negative compartment signal represent the B compartment. For HiCExplorer software, the parameter "--extraTrack" in the hicPCA command could be used for this work. Bed files interpreted as gene tracks should be provided.

3.3.7 Identification of Significant Interactions

For prokaryotes, we recommend using the Fit-Hi-C software [29] to calculate significant interactions. The resolution can be set to 2 kb or 5 kb. Filtering thresholds can be applied with a *P*-value and *Q*-value less than 0.01 and contact counts greater than 2 (*see* **Note 17**). Three files should be prepared: the first is the Hi-C read depth of each bin (input.fragment), the second is the raw contacts between bins (input.interaction), and the third is the bias information file generated by balancing (input.bias). Using a.cool file as input, the cooler software commands "cooler dump input.cool -o output.matrix" and "cooler dump -t chroms input.cool -o output. bias" can be used to extract the contacts and bias information, which can then be formatted for software input as described on the Fit-Hi-C website (https://github.com/ay-lab/fithic). To run Fit-Hi-C, use the parameters "-f input.fragment -i input.interaction -t input.bias -p 2 -r resolution -o outpath." The output of Fit-Hi-C is a gzip-compressed txt file containing interaction information, with the *P*-value in column 6 and the *Q*-value in column 7.

4 Notes

1. Ensure that the flasks used for *Streptomyces* culture have been thoroughly cleaned, as even traces of detergents like SDS can inhibit vegetative cell growth.

2. To avoid cross-contamination when handling spores of different strains, dispense 10 mL aliquots of 20% glycerol into separate 15 mL tubes. Store these aliquots at RT for future use.
3. Always start from a single colony. Propagation through successive rounds of mass culture is discouraged.
4. Approximately 7×10^9 spores can be harvested from a fully sporulated petri dish. The recommended concentration for frozen spores is 1×10^9 spores/mL, with aliquots stored for future experimental use. For reproducibility, spores from the same batch should be used across experiments.
5. While *Streptomyces* spores retain viability through repeated freeze-thaw cycles, prolonged exposure to elevated temperatures can induce germination. Germinated spores do not survive freezing, compromising sample quality. To maintain integrity, it is advisable to avoid multiple freeze-thaw cycles.
6. For physiological research, uniformity in growth is indispensable. *Streptomyces* tend to form compact clumps in liquid culture, with significant differences in gene expression and metabolism between cells at the center and the periphery of these aggregates. This heterogeneity can lead to capturing mixed cell states, complicating the interpretation of 3D chromatin structures. Achieving dispersed growth is therefore essential to ensure that Hi-C analysis accurately reflects a single physiological state. A key strategy for achieving optimal dispersion involves using a large inoculum and applying a pre-treatment that promotes spore germination without inducing excessive outgrowth. This protocol is optimized for *Streptomyces coelicolor*, but modifications may be required for other strains. For further guidance, refer to Chap. 2, "Growth and Preservation of *Streptomyces*" in *Practical Streptomyces Genetics* [2].
7. Accurate cell counting for *Streptomyces* mycelium is challenging. After centrifugation, each tube will contain approximately 200 μL of the original culture ($OD_{450} = 2$), corresponding to roughly 2×10^8 cells. Using too many cells can interfere with lysozyme digestion, while too few may result in inefficient DNA extraction.
8. The supernatant contains mostly non-interacting naked DNA, which does not contribute significantly to interaction profiling. Removing it by centrifugation helps improve the signal-to-noise ratio in the Hi-C results.
9. It is recommended to use Buffer AE for elution, as using ddH_2O with a pH below 8.0 can significantly reduce elution efficiency and lead to a lower DNA yield.
10. The total amount of DNA should be more than 1 μg.

11. The duration of the 37 °C incubation can be adjusted based on the desired DNA fragment size. Longer incubations will result in smaller fragment sizes.
12. The DNA remains bound to the beads in this step.
13. For adapter sources other than VAHTS, use a working concentration of 10 μM.
14. The number of cycles depends on the amount of target sequencing data needed. For prokaryotic samples, 9 cycles are sufficient because their genomes are smaller and less complex, requiring less amplification to achieve good coverage while avoiding bias.
15. Different window size selection can lead to significantly different results. A larger window size may generate larger structural domains, while a smaller window size may generate smaller domains.
16. Few studies in prokaryotic 3D genomics have explored A/B compartmentalization, but our previous research suggests that *Streptomyces* may exhibit this level of structure.
17. For circular prokaryotic genomes, the genome distance calculation code in the Fit-Hi-C software should be modified. Fit-Hi-C assumes a linear genome, where distances are measured directly along the sequence. However, for circular genomes, interaction distances need adjustment to reflect the shortest path around the circular structure. Specifically, if the interaction distance is less than half the chromosome length, it remains unchanged. If it exceeds half the chromosome length, the distance should be subtracted from the total genome length to represent the correct shorter path.

Acknowledgments

This work was supported by the National Key R&D Program of China (2018YFA0903200) to Y. S.

References

1. Hopwood DA (2007) Streptomyces in nature and medicine: the antibiotic makers. Oxford University Press, New York
2. Kieser T (2000) Practical Streptomyces genetics. The John Innes Foundation, Norwich
3. Muteeb G, Rehman MT, Shahwan M, Aatif M (2023) Origin of antibiotics and antibiotic resistance, and their impacts on drug development: a narrative review. Pharmaceuticals (Basel) 16:1615
4. Kalkreuter E, Pan G, Cepeda AJ, Shen B (2020) Targeting bacterial genomes for natural product discovery. Trends Pharmacol Sci 41: 13–26
5. Bryant JA, Sellars LE, Busby SJW, Lee DJ (2014) Chromosome position effects on gene expression in *Escherichia coli* K-12. Nucleic Acids Res 42:11383–11392
6. Bilyk B, Horbal L, Luzhetskyy A (2017) Chromosomal position effect influences the

heterologous expression of genes and biosynthetic gene clusters in *Streptomyces albus* J1074. Microb Cell Factories 16:5

7. Dekker J, Rippe K, Dekker M, Kleckner N (2002) Capturing chromosome conformation. Science 295:1306–1311
8. Zhao Z, Tavoosidana G, Sjölinder M et al (2006) Circular chromosome conformation capture (4C) uncovers extensive networks of epigenetically regulated intra- and interchromosomal interactions. Nat Genet 38:1341–1347
9. Lieberman-Aiden E, Van Berkum NL, Williams L et al (2009) Comprehensive mapping of long-range interactions reveals folding principles of the human genome. Science 326:289–293
10. Szafran MJ, Małecki T, Strzałka A et al (2021) Spatial rearrangement of the *Streptomyces venezuelae* linear chromosome during sporogenic development. Nat Commun 12:5222
11. Lioy VS, Lorenzi J-N, Najah S et al (2021) Dynamics of the compartmentalized *Streptomyces* chromosome during metabolic differentiation. Nat Commun 12:5221
12. Deng L, Zhao Z, Liu L et al (2023) Dissection of 3D chromosome organization in *Streptomyces coelicolor* A3(2) leads to biosynthetic gene cluster overexpression. Proc Natl Acad Sci USA 20:e2222045120
13. Shao W, Wang J, Zhang Y et al (2024) The jet-like chromatin structure defines active secondary metabolism in fungi. Nucleic Acids Res 52:4906–4921
14. Le TBK (2019) Chromosome conformation capture with deep sequencing to study the roles of the structural maintenance of chromosomes complex in vivo. Methods Mol Biol 2004:105–118
15. Bolger AM, Lohse M, Usadel B (2014) Trimmomatic: a flexible trimmer for Illumina sequence data. Bioinformatics 30:2114–2120
16. Wingett S, Ewels P, Furlan-Magaril M et al (2015) HiCUP: pipeline for mapping and processing Hi-C data. F1000Res 4:1310
17. Servant N, Varoquaux N, Lajoie BR et al (2015) HiC-Pro: an optimized and flexible pipeline for Hi-C data processing. Genome Biol 16:259
18. Zhang H, Song L, Wang X et al (2021) Fast alignment and preprocessing of chromatin profiles with Chromap. Nat Commun 12:6566
19. Li H (2013) Aligning sequence reads, clone sequences and assembly contigs with BWA-MEM. arXiv. https://arxiv.org/abs/1303.3997v2
20. Abdennur N, Mirny LA (2020) Cooler: scalable storage for Hi-C data and other genomically labeled arrays. Bioinformatics 36:311–316
21. Schmitt AD, Hu M, Jung I et al (2016) A compendium of chromatin contact maps reveals spatially active regions in the human genome. Cell Rep 17:2042–2059
22. Lioy VS, Cournac A, Marbouty M et al (2018) Multiscale structuring of the *E. coli* chromosome by nucleoid-associated and condensin proteins. Cell 172:771–783.e18
23. Ramírez F, Bhardwaj V, Arrigoni L et al (2018) High-resolution TADs reveal DNA sequences underlying genome organization in flies. Nat Commun 9:189
24. Le TBK, Imakaev MV, Mirny LA, Laub MT (2013) High-resolution mapping of the spatial organization of a bacterial chromosome. Science 342:731–734
25. Varoquaux N, Ay F, Noble WS, Vert J-P (2014) A statistical approach for inferring the 3D structure of the genome. Bioinformatics 30:i26–i33
26. Rieber L, Mahony S (2017) miniMDS: 3D structural inference from high-resolution Hi-C data. Bioinformatics 33:i261–i266
27. Schrödinger L, DeLano W (2020) PyMOL. Retrieved from http://www.pymol.org/pymol
28. Zheng X, Zheng Y (2018) CscoreTool: fast Hi-C compartment analysis at high resolution. Bioinformatics 34:1568–1570
29. Ay F, Bailey TL, Noble WS (2014) Statistical confidence estimation for Hi-C data reveals regulatory chromatin contacts. Genome Res 24:999–1011

Chapter 14

Mapping the 3D Chromosome Organization of a Biosynthetic Gene Cluster by Capture Hi-C (CHi-C)

George Jervis and Hans-Wilhelm Nützmann

Abstract

Chromosome conformation capture (3C), when combined with high-throughput next-generation sequencing, known as Hi-C, is a powerful approach for visualizing three-dimensional genome architecture. This technique enables detailed study of how chromosome topology influences gene regulation and other key cellular processes.

Incorporating a sequence capture step into the Hi-C workflow enables targeted enrichment of predefined genomic regions of interest. Compared with conventional Hi-C, Capture Hi-C (CHi-C) increases the sequencing depth for the targeted regions, enabling high-resolution analysis of chromatin structure within a predefined locus at a lower cost.

Here, we present a detailed protocol for CHi-C applied to a region of the *Arabidopsis thaliana* genome encompassing the thalianol biosynthetic gene cluster.

Key words Capture Hi-C, *Arabidopsis thaliana*, Chromosome conformation capture, Sequence-capture, Biosynthetic gene cluster

1 Introduction

Three-dimensional genome organization plays a critical role in the dynamic control of fundamental process such as DNA replication, gene expression, and cell differentiation [1–3].

The use of chromosome confirmation capture (3C) techniques enables the characterization of chromosomal interactions in the nuclear space [1, 4]. High-throughput chromosome conformation capture (Hi-C) which integrates 3C with next-generation sequencing, enables genome-wide analysis of chromatin interactions [5]. The Hi-C technique is well established in *Arabidopsis thaliana* and other plant species including important crops [6–8].

However, obtaining high-resolution genome-wide Hi-C coverage remains costly due to the sequencing depth required. To address this limitation, sequence-capture technology can be integrated into Hi-C protocols to enrich specific genomic loci of

Hans-Wilhelm Nützmann and Ramesha Thimmappa (eds.), *Metabolic Gene Clusters: Methods and Protocols*,
Methods in Molecular Biology, vol. 3054, https://doi.org/10.1007/978-1-0716-5384-5_14,

interest (Capture Hi-C). In this approach, custom-designed biotinylated RNA probes hybridize with target regions within a Hi-C library. The hybridized fragments are then isolated using streptavidin-coated beads, while non-target sequences are washed away. Amplification of the captured library produces a Hi-C dataset enriched for the selected genomic regions, enabling higher-resolution analysis at reduced sequencing cost [9].

Capture Hi-C enables the high-resolution visualization of topology of specific genomic loci, such as the thalianol biosynthetic gene cluster (BGC). BGCs are functionally related, colocalized, and coregulated genes that together encode specialized metabolic pathways. The thalianol BGC, for example, contributes to the biosynthesis of root-specific metabolites that play a key role in shaping the root microbiome of *Arabidopsis thaliana* [10]. Using Capture Hi-C, we have previously demonstrated the thalianol BGC resides in a dynamic chromosomal domain that undergoes organ-dependent changes in local conformation and nuclear positioning [11].

In this chapter, we present an optimized step-by-step protocol for generating a Capture Hi-C library enriched to a specific region of the *Arabidopsis thaliana* genome. This approach enables high-resolution visualization of chromosome topology at the targeted loci. The sequence capture component of this protocol can be readily adapted to target any region of interest in the *Arabidopsis thaliana* genome and modified for application in Hi-C protocols of other plant species.

2 Materials

Prepare all solutions using ultrapure nuclease-free water unless stated otherwise. Store all reagents at temperature recommended by manufacture unless indicated otherwise. Follow all relevant waste disposal regulations when disposing waste materials.

1. Plant growth chamber.
2. 1 g of *Arabidopsis thaliana* seeds.
3. 100 mm Square petri dishes.
4. MS Phytagel (Murashige & Skoog Media 4.3 g per liter, Sucrose 10 g per liter, Phytagel 5 g per liter, 5.8 pH with NaOH) Pour immediately into square petri dishes following autoclaving.
5. Fume hood.
6. Vacuum pump and chamber.
7. Thermal block with shaking function.
8. Benchtop refrigerated centrifuge for 50 mL tubes.

9. Benchtop refrigerated centrifuge for 2 mL tubes.
10. Microcentrifuge.
11. 50 mL Falcon tubes.
12. 1.5 mL LoBind Eppendorf tubes.
13. 2 mL LoBind Eppendorf tubes.
14. 0.2 mL PCR tubes.
15. Miracloth (Millipore).
16. Nylon net filter (30 μM pore).
17. Funnels.
18. Plastic mesh.
19. Scissors.
20. Multi-well low-profile reagent reservoirs.
21. Razor blades.
22. Neubauer counting chamber.
23. Epifluorescence microscope.
24. Nuclei Isolation Buffer (NIB): 20 mM HEPES, pH 8.0, 250 mM sucrose, 1 mM $MgCl_2$, 5 mM KCl, 40% (v/v) glycerol, 0.25% (v/v) Triton X-100, 0.1 mM PMSF, and 0.1% (v/v) β-mercaptoethanol. Stir thoroughly using a magnetic stirrer and keep ice-cold (*see* **Note 1**).
25. 1 × PBS.
26. 37% formaldehyde solution.
27. 2 M Glycine solution (made fresh in nuclease-free water).
28. DAPI (1 mg/mL).
29. Qubit 4 Fluorometer with 1× dsDNA HS Assay Kit.
30. Agilent 4150 TapeStation System with D1000 ScreenTape.
31. Bioruptor Pico.
32. Bioruptor Pico 1.5 mL Microtubes with caps.
33. PCR Thermocycler (×2).
34. DynaMag for 0.2 mL tubes.
35. Nutator.
36. 20% SDS Ultrapure.
37. 10% Triton-X-100.
38. DpnII restriction enzyme.
39. 10× NEB r3.1 Buffer.
40. Biotin-14-dATP.
41. DNA Polymerase I, Large (Klenow) Fragment.
42. T4 DNA Ligase.

43. 10× T4 DNA Ligase Buffer.
44. Bovine Serum Albumin (BSA) (10 mg/mL).
45. 25:24:1 (v/v/v) phenol-chloroform-isoamyl alcohol.
46. 3 M Sodium Acetate (pH 5.2).
47. Glycogen (20 mg/mL).
48. Ethanol.
49. RNAse A (20 mg/mL).
50. Agarose.
51. GeneRuler 1Kb Plus DNA Ladder.
52. SYBR Safe DNA Gel Stain.
53. Agarose gel casting tray and combs.
54. DNA electrophoresis chamber and power supply.
55. T4 DNA Polymerase.
56. 10× NEB 2.1 Buffer.
57. 0.1× TE Buffer: 10 mM Tris-HCl (pH 7.5), 0.1 mM EDTA.
58. 2× Binding and Washing (B&W) buffer: 10 mM Tris-HCl (pH 7.5), 1 mM EDTA, 2 M NaCl.
59. Dynabeads MyOne Streptavidin C1.
60. NEBNext Ultra II DNA Library Prep Kit for Illumina.
61. NEBNext Multiplex Oligos for Illumina.
62. Beckman Coulter SPRIselect Beads.
63. Genevac EZ-2 4.0 Series Centrifugal Evaporator.
64. SureSelect Max Target Fast Hyb Kit.
65. SureSelect Max Blockers and Primers Kit.
66. SureSelect Max Purification Beads.
67. SureSelect HS probe (Designed and ordered for specific region capture).

3 Methods

3.1 Hi-C

In this section, we will cover the preparation of a Hi-C library. Briefly, this includes the fixation and isolation of *Arabidopsis thaliana* nuclei, digestion of the chromatin into fragments with a restriction enzyme, the labelling of fragment overhangs with biotinylated nucleotides, ligation of the biotinylated fragments, purification of the ligation products, and the conversion of these purified fragments to a library that can be sequenced.

The initial steps of the protocol up to and including DNA extraction (Subheading 3.1.10) take 4 consecutive days. Afterwards, the library can be stored at −20 °C for prolonged periods of time.

This protocol is optimized for the preparation of a single Hi-C library, with the option to produce multiple libraries in parallel.

3.1.1 Preparation of Plant Material

1. Surface sterilize 1 g of *Arabidopsis thaliana* seeds (*see* **Note 2**). Store in cold dark environment for 3–4 days to ensure synchronous germination.
2. Following cold storage, sow two rows of seeds onto 20 MS phytagel 100 mm square plates, ensuring that seeds are spread out with room to grow. Store plates vertically in plant growth chamber for 7 days under long-day conditions (16 h light/8 h dark, 22 °C) (*see* **Note 3**).

3.1.2 Fixation of Plant Tissue

1. Quickly remove roots from 7-day-old seedlings with a razor blade and submerge the material in a pre-cooled 50 mL Falcon tube containing 35 mL of ice-cold 1 × PBS.
2. Add 2 mL of 37% formaldehyde solution to the 1 × PBS + plant material (2% final concentration) under the fume hood. It is critical to keep time short between cutting and submerging the plant material in formaldehyde solution.
3. Vacuum infiltrate the plant material for 45 min under a fume hood. Interrupt the vacuum pump every 15 min and give the Falcon tube a shake to encourage fixative to enter the plant tissue.
4. To stop cross-linking reaction, add 2.5 mL of fresh 2 M glycine to a final concentration of 0.125 M. Mix thoroughly by inverting the Falcon tube several times and perform another 5 min of vacuum infiltration.
5. Following infiltration, add hard mesh to the Falcon tube. Using the mesh to trap the plant material and carefully pour away the formaldehyde solution into a suitable waste container. Replace the solution with ice-cold 1 × PBS.

3.1.3 Nuclei Isolation

1. After four separate washes in ice-cold 1 × PBS, quickly remove the plant material, pat dry, and place in a multi-well low-profile reagent reservoir kept on ice. Pour in ice-cold NIB, completely immersing the plant material. Put 50 mL Falcon tubes on ice at this point.
2. Finely chop the plant material with a fresh razor blade in NIB for at least 10 min until solution is slurry-like.
3. Carefully transfer the plant slurry through a layer of Miracloth placed in a funnel into an ice-cold 50 mL Falcon tube. Filter the resulting suspension through a 30 μM pore size nylon net mesh into a second ice-cold 50 mL Falcon tube (*see* **Notes 4** and **5**).
4. Centrifuge the 50 mL Falcon tube containing the nuclei solution for 15 min (4 °C, 3000 g) until a pellet has formed.

Carefully remove the supernatant, then gently resuspend the nuclei pellet in 1 mL of NIB and transfer the solution into a 1.5 mL LoBind Eppendorf tube.

5. Centrifuge the 1.5 mL tube for 5 min (4 °C, 1900 g), remove the supernatant, and resuspend the pellet in NIB. Repeat this wash once more in the same 1.5 mL Eppendorf tube.
6. To assess nuclei quality and quantity, take a 10 μL aliquot for nuclei counting. Stain the aliquoted nuclei with DAPI and count them using a Neubauer counting chamber under an epifluorescence microscope. Proceed with approximately 10^6–10^7 nuclei for subsequent steps (*see* **Note 6**).
7. Prepare 2 mL of 1.2× NEB r3.1 buffer and keep on ice.
8. Wash the nuclei three times with ice-cold 1.2× NEB r3.1 buffer using consecutive centrifugation steps at 1900 g and 4 °C for 5 min each. After the final wash, resuspend the pellet in 500 μL of 1.2× NEB r3.1 buffer.

3.1.4 Nuclei Permeabilization

1. To inactivate endogenous enzymes and permeabilize the nuclei membrane, add 7.6 μL of 20% SDS to the nuclei solution (0.3% final concentration).
2. Incubate the sample at 65 °C for 25 min at 900 rpm, then at 37 °C for 15 min at 900 rpm.
3. To sequester the SDS, add 111.6 μL of fresh 10% Triton-X-100 (1.8% final concentration) to the sample and incubate at 37 °C for 1 h at 900 rpm.
4. At this point, remove a 5% aliquot of the sample (30 μL), transfer to a 1.5 mL LoBind Eppendorf tube and store at −20 °C. This sample will be called the pre-digestion control and will be used to test the quality of DNA and check for degradation.

3.1.5 Restriction Enzyme Digestion

1. Add 400 U of high concentration DpnII enzyme to the remaining sample and incubate overnight at 37 °C 900 rpm (*see* **Note 7**).
2. The following morning, inactivate the restriction enzyme by incubating sample at 65 °C for 20 min. Place sample on ice immediately following incubation.

3.1.6 Biotinylation of Overhangs

1. Prepare and add the following mix to the sample and incubate at 23 °C for 4 h. Add Klenow fragment separately from the rest of the mix. (4 μL 10× NEB r3.1, 6.5 μL nuclease-free water, 1.5 μL 10 mM dCTP, 1.5 μL 10 mM dGTP, 1.5 μL 10 mM TTP, 15 μL 1 mM biotin-14-dATP, 10 μL DNA Polymerase 1, Large (Klenow) Fragment (5 U/μL)). Gently mix by pipetting up and down every 30 min.

3.1.7 Ligation

1. Transfer the sample from 1.5 mL tube to a 2 mL LoBind Eppendorf tube.
2. Dilute the biotinylated sample with 200 μL 10× T4 DNA ligation buffer, 220 μL 10% Triton-X-100 (1.1% final concentration), and nuclease-free water to reach a final volume of 2 mL.
3. Prior to adding ligase, take another 5% aliquot (100 μL) to assess the efficiency of the DpnII digestion. Transfer to a 1.5 mL LoBind Eppendorf tube and store at −20 °C. This sample will be called the pre-ligation control.
4. Add 200 U of T4 DNA Ligase to the sample and mix by inverting the tube several times.
5. Incubate the sample for 5 h or overnight at 16 °C (*see* **Note 8**).

3.1.8 Reversal of Cross-Linking

1. Following ligation, add 15 μL of Proteinase K (20 mg/mL) to the Hi-C sample, mix via pipetting up and down, and incubate overnight at 65 °C.
2. For the pre-digestion and pre-ligation controls, remove samples from the freezer and dilute to 500 μL with nuclease-free water. Add 2.5 μL of Proteinase K to controls and incubate in parallel to Hi-C sample, at 65 °C overnight.
3. The following morning, add 30 μL of RNAse A (10 mg/mL) to Hi-C sample, and add 5 μL to pre-digestion and pre-ligation controls. Incubate all samples at 37 °C for 40 min. Immediately place samples on ice following incubation.

3.1.9 DNA Extraction

1. Take phenol-chloroform-isoamyl alcohol out of the fridge and allow to reach room temperature to improve phase separation.
2. Split the Hi-C sample into four 1.5 mL LoBind Eppendorf tubes (500 μL each), keep the control samples in their original tubes. Allow samples to reach room temperature. Add an equal volume of phenol-chloroform-isoamyl alcohol to all samples and shake vigorously.
3. Centrifuge tubes for 10 min at 4500 g. Carefully transfer DNA-containing aqueous upper phase to a new 1.5 mL LoBind Eppendorf tube.
4. Precipitate DNA in each of the 500 μL samples by adding 50 μL 3 M sodium acetate (pH 5.2). 2 μL glycogen and 1100 μL ice-cold 100% ethanol.
5. Mix by inverting several times and incubate at −80 °C for 1 h, then move samples to −20 °C for 1 h. Centrifuge at 20,000 g for 1 h at 4 °C.
6. Gently remove the supernatant, leaving a small pellet of DNA at the bottom of each sample. This pellet is then washed in ice-cold 70% ethanol. Centrifuge the tubes at 20,000 g for 10 min at 4 °C and remove the supernatant.

7. Allow the DNA pellets to air-dry at room temperature for approximately 5 min. Avoid over-drying, as this may hinder efficient resuspension.
8. Dissolve the Hi-C sample pellets in 50 μL of nuclease-free water and combine for a total of 200 μL. Resuspend the pre-digestion and pre-ligation controls in 20 μL of nuclease-free water.
9. Store samples in the fridge overnight at 4 °C.

3.1.10 Testing Sample Quality

It is now possible to assess DNA quality, digestion, and ligation efficiency of the Hi-C sample.

1. Run the Hi-C sample, pre-digestion, and pre-ligation control, with a 1 kb + DNA ladder via gel electrophoresis (0.8% agarose gel, SYBR stained) side by side using a 10 μL aliquot sample of each. The pre-digestion control should display a single high-molecular-weight band. The pre-ligation control is expected to show a DNA smear, indicative of successful restriction enzyme digestion. The Hi-C sample should appear as a higher molecular weight smear compared to the pre-ligation control, reflecting successful proximity ligation (*see* **Note 9**).
2. Quantify the concentration of the Hi-C sample via Qubit Fluorometer with dsDNA HS Assay Kit. A Hi-C sample prepared from *Arabidopsis thaliana* roots harvested across 20 plates should yield approximately 3000 ng of DNA.

3.1.11 Biotin Removal

1. After diluting the total Hi-C sample to 500 μL in nuclease-free water, a second DNA extraction was performed to the same instructions as Subheading 3.1.9, resuspending the pellet in 50 μL of nuclease free water. To remove biotinylated dCTP from unligated DNA ends, add the following mix to the 50 μL ligated Hi-C sample. 12 μL 10× NEB 2.1 Buffer, 0.3 μL 10 mM ATP, 0.3 μL 10 mM GTP, 0.6 μL BSA, 5 μL T4 DNA Polymerase (3000 U/mL), and 51.8 μL nuclease-free water to reach 120 μL. Incubate the sample at 20 °C for 4 h.
2. Inactivate the reaction via a 20-min treatment at 75 °C.
3. Dilute the sample to 500 μL in nuclease free water and perform a third DNA extraction to the same instructions as Subheading 3.1.9. Resuspend pellet to 50 μL in 0.1× TE buffer and measure concentration via Qubit. Keep sample in fridge at 4 °C.

3.1.12 DNA Shearing

1. Dilute the Hi-C sample to 300 μL in 0.1× TE Buffer and transfer to a Bioruptor 1.5 mL microtube with cap (*see* **Note 10**).
2. Take a 1 μL aliquot as a pre-sonication control.

3. Briefly vortex and centrifuge the sample. Incubate sample on ice for 10 min.
4. Shear sample to 200–400 bp length using a Diagenode Bioruptor Pico (6 cycles, 30 s on/30 s off). Make sure to keep samples ice-cold to ensure effective sonication.
5. Transfer post-sonication sample to a 1.5 mL LoBind Eppendorf and dilute to 500 μL in 0.1× TE Buffer. Allow sample to reach room temperature.
6. Perform a fourth DNA extraction to the same instructions as Subheading 3.1.9. Resuspend pellet to 50 μL in 0.1× TE. Keep sample in fridge at 4 °C.
7. Run sample and pre-sonication control on an Agilent 4150 TapeStation System with D1000 ScreenTape to analyze effectiveness of sonication, indicated by the shearing of larger fragments down to 200–400 BP.

3.1.13 Library Preparation, End Prep, and Adapter Ligation

The next steps of library preparation are following the NEBNext Ultra II DNA Library Prep Kit for Illumina, and NEBNext Multiplex Oligos for Illumina. The NEBNext Ultra II kit recommends 500 pg–1 μg of sheared DNA per sample.

1. Transfer 50 μL of Hi-C sample to a 0.2 mL PCR tube and keep on ice. Add 7 μL NEBNext Ultra II End Prep Reaction Buffer and 3 μL NEBNext Ultra II End Prep Enzyme Mix. Pipette the mix slowly up and down at least 10 times. Perform a quick spin to collect all liquid from the sides of the tube.
2. Place the tube into a thermocycler, with the heated lid set to ≥75 °C and run for 30 min at 20 °C, then 30 min at 65 °C, and hold at 4 °C.
3. Perform adaptor ligation by adding the following separately to the 60 μL sample. 2.5 μL 15 μM NEBNext Adaptor for Illumina, 30 μL NEBNext Ultra II Ligation Master Mix, and 1 μL NEBNext Ligation Enhancer to a total volume of 93.5 μL. Pipette the entire volume of this solution up and down at least 10 times to mix thoroughly. The Ligation Master mix is viscous; take care to ensure adequate mixing of the reaction. Perform a quick spin to collect all liquid from sides of the tube (*see* **Note 11**).
4. Incubate this mix for 15 min at 20 °C in a thermocycler with the heated lid off.
5. Add 3 μL of USER enzyme to the ligation mixture. Mix the solution well by pipetting up and down; incubate at 37 °C for 15 min with the heated lid set to ≥47 °C. Place sample on ice following.
6. The Hi-C sample can be stored overnight at 4 °C or −20 °C for prolonged storage.

3.1.14 Size Selection

To clean-up sample and narrow the library size range, the sample requires size selection between 200 and 650 bp using a double-sided 0.85–0.61 (Left-Right) SPRIselect bead selection.

1. Add 3.5 μL of 0.1× TE solution to bring the sample volume to 100 μL and let the sample reach room temperature.
2. Vigorously shake the SPRIselect bottle to resuspend the beads. Let the beads equilibrate to room temperature before use.
3. For left-handed size selection, add 85 μL of SPRIselect beads to the sample. Mix the reaction volume by pipetting 10 times then incubate at room temperature for 1 min.
4. Place the 0.2 mL tube on the DynaMag and allow a few minutes for the SPRI beads to settle to the magnet. Carefully remove and discard the supernatant, avoid removing any beads.
5. With the tube still on the magnet, add 180 μL of fresh 85% ethanol and incubate at room temperature for 30 s. Carefully remove and discard the ethanol supernatant, avoid disrupting the beads.
6. To elute the sample, remove the tube from the DynaMag and add 100 μL of nuclease-free water. Resuspend the beads by pipetting up and down 10 times, then incubate the reaction for 1 min at room temperature.
7. Place the tube on the DynaMag and allow the beads to settle on the magnet for a few minutes. Remove the size-selected supernatant and place the sample into new 0.2 mL tube.
8. To perform the right-sided selection, add 61 μL of SPRIselect beads to the 100 μL sample. Mix the reaction volume by pipetting up and down 10 times, then incubate at room temperature for 1 min.
9. Place the sample on the DynaMag and allow the SPRI beads to settle against the magnet for several minutes. Carefully transfer the supernatant to a new 0.2 mL tube, avoid carryover of any beads. Discard the tube with the remaining beads.
10. Add 119 μL of SPRIselect beads to the transferred supernatant. Mix thoroughly by pipetting up and down 10 times, then incubate for 1 min at room temperature.
11. Place the 0.2 mL tubes on the DynaMag and allow the SPRI beads to settle against the magnet for several minutes. Carefully remove and discard the supernatant, avoiding the disturbance of any beads.
12. With the tubes still on the magnet, add 180 μL of fresh 85% ethanol and incubate at room temperature for 30 s. Carefully remove and discard the ethanol supernatant without disrupting the beads.

13. To elute the sample, remove the tube from the magnet and add 50 μL of nuclease-free water. Resuspend the beads by pipetting up and down 10 times, then incubate for 1 min at room temperature.
14. Place the tube back on the DynaMag and allow the beads to settle on the magnet for several minutes. Transfer the size-selected supernatant to a new 0.2 mL tube.
15. Store the sample at 4 °C overnight or −20 °C for prolonged storage.
16. Quantify the sample concentration and assess the size distribution using the Qubit fluorometer and TapeStation, respectively, to evaluate the effectiveness of the size selection.

3.1.15 Biotin Pulldown

To remove unwanted non-biotinylated fragments, a biotin-streptavidin bead pulldown is performed using Dynabeads MyOne Streptavidin C1 beads.

1. Prior to the starting the pulldown, freshly prepare 2× Binding and Washing (B&W) buffer in nuclease-free water.
2. Take Dynabeads MyOne Streptavidin C1 bottle out of the fridge and let the beads equilibrate to room temperature for 30 min.
3. Vigorously shake the Dynabeads MyOne Streptavidin C1 bottle to resuspend the beads.
4. Transfer 25 μL of beads to a 0.2 mL tube. Add 25 μL of 1× B&W (Diluted with an equal volume of nuclease-free water) and resuspend the beads.
5. Place the tubes on the DynaMag and allow the beads to settle for 1 min. Carefully remove and discard the supernatant.
6. Remove the tube from the DynaMag and resuspend the washed beads in 25 μL of 1× B&W buffer.
7. Place the tubes on the DynaMag and allow the beads to settle for 1 min. Carefully remove and discard the supernatant. Repeat this washing step two more times.
8. Resuspend the Dynabeads in 50 μL of 2× B&W solution (5 μg/μL bead concentration).
9. Add 50 μL of Hi-C sample to the Dynabead solution. Optimal binding occurs when NaCl concentration in buffer is reduced from 2 M to 1 M.
10. Place the sample-bead solution on a nutator and incubate the sample at room temperature for 15 min with gentle rotation.
11. After incubation, place tubes on DynaMag rack and allow 2–3 min for beads to settle on the magnet. Carefully remove the supernatant and discard.

12. Wash the coated beads 3 times with 100 μL of 1× B&W Buffer.
13. Resuspend the beads in 50 μL of nuclease-free water and store in the fridge at 4 °C (*see* **Note 12**).

3.1.16 Library Amplification

To amplify a Hi-C library from the pulldown beads, set up a PCR reaction using NEBNext Ultra II DNA Library Prep Kit for Illumina in combination with NEBNext Multiplex Oligos for Illumina.

1. Transfer 15 μL of Hi-C bead solution to 0.2 mL tube, add 25 μL 2× NEB Next Ultra II Q5 Master Mix, 5 μL 10 μM NEBNext Universal PCR i5 Primer and 5 μL 10 μM NEBNext Index i7 Primer (sample specific). Mix solution thoroughly by pipetting up and down at least 10 times. Perform a quick spin to collect the liquid from the sides of the tube (*see* **Note 13**).
2. Place tube in thermocycler and run the following program: 1 cycle: 30 s 98 °C (initial denaturation); 10 cycles: 10 s 98 °C (denaturation), 75 s 65 °C (annealing/extension); 1 cycle: 5 min 65 °C (Final Extension); hold at 4 °C. The number of cycles can be optimized based on input amount and desired yield.
3. Transfer the 0.2 mL tube to the DynaMag and allow beads to settle for a few minutes. Carefully transfer supernatant containing amplified Hi-C library to a new 0.2 mL tube. Keep sample at 4 °C or −20 °C for prolonged storage (*see* **Note 14**).

3.1.17 Library Cleanup

Amplified Hi-C library is purified via Agencourt AMPure XP PCR purification.

1. Remove Agencourt AMPure XP bottle from the fridge and allow it to equilibrate to room temperature for at least 30 min. Vortex or shake bottle vigorously to fully resuspend beads.
2. Add 90 μL of Ampure XP beads to 50 μL of amplified Hi-C sample. Mix reaction thoroughly by pipetting up and down 10 times. Incubate for 5 min at room temperature for maximum recovery.
3. Place tube on DynaMag rack and allow beads to settle on the magnet for a few minutes. Carefully remove and discard the supernatant. It is recommended to leave a small volume of supernatant behind to avoid removing beads.
4. With the tube still on the DynaMag rack, add 180 μL of 70% ethanol to the beads and incubate for 30 s at room temperature. Remove and discard the ethanol solution. Repeat this for two more washes.
5. Remove all ethanol solution from beads and remove tube from the rack. Allow the beads to air-dry for a minute to remove any

residual ethanol solution. Be careful not to over dry the beads (beads appear cracked if overdried) as this will decrease elution efficiency.

6. Add 50 μL of nuclease-free water to the beads and resuspend by pipetting the mix up and down 10 times. Incubate the solution at room temperature for 2 min.
7. Place the reaction on the DynaMag rack and let the beads settle on the magnet for a couple of minutes. Carefully remove and transfer the purified Hi-C library solution to a new 0.2 mL tube. Store sample at 4 °C or −20 °C for prolonged periods.

3.2 Targeted Sequence Capture

This targeted sequence capture protocol for pooled Hi-C libraries is modified from the Agilent SureSelect Max Target Enrichment using Fast Hybridization protocol. It incorporates the use of SureSelect Fast Hyb Kit, SureSelect Max Blockers and Primers Kit, SureSelect Max Purification Beads, and SureSelect XT HS Probes.

Briefly, custom-designed biotinylated RNA probes are hybridized with the chosen Hi-C libraries. These hybridized fragments are then captured using streptavidin-coated magnetic beads, allowing non-targeted regions of the library to be washed away. The bead-bound library is subsequently amplified by PCR to enrich those selected regions.

3.2.1 Planning and Probe Design

1. To capture and enrich a specific region of interest, design probes in accordance with Agilent SureSelect workflow.
2. Using a chosen genome assembly, select a region of interest and design 120 bp probes to cover all DpnII restriction fragments within. For each fragment longer than 250 bp, design two probes positioned near the ends of the fragment. For fragments shorter than 250 bp, select a single centrally located probe. Exclude fragments that are shorter than 120 bp, as these are too short to accommodate a probe. In all cases, choose the most optimal probe in terms of sequence uniqueness and GC content (Fig. 1) (*see* **Note 15**).

3.2.2 Library Pooling

1. To save time and reagents, multiple Hi-C libraries can be pooled into a single capture reaction. Up to 16 libraries can be pooled to a total of 4 μg DNA (250 ng per library) into a

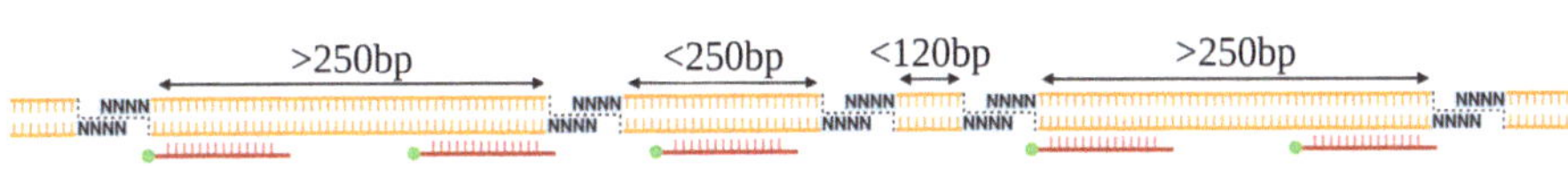

Fig. 1 Sequence capture probe design. A genomic region is shown divided by restriction enzyme sites. Red single-strand probes represent the optimal positions for 120 bp biotinylated RNA probe design. For fragment longer than 250 bp, two probes are placed near each end. For fragments shorter than 250 bp, a single central probe is selected. Fragments shorter than 120 bp are too short to accommodate a probe [12]

0.2 mL tube. When pooling multiple libraries, ensure each library has a unique i7 index primer for demultiplexing during sequencing.

2. Concentrate the pooled library solution using a Genevac Evaporator, held to ≤45 °C. Without completely drying the sample, reduce the library pool volume to below 12 μL. Measure sample and adjust to 12 μL volume using nuclease-free water.
3. Vortex the tube vigorously for 10 s then briefly spin the tube to collect all liquid. Keep samples on ice.

3.2.3 Hybridize Libraries to SureSelect Probes

1. Preset a thermocycler with heated lid set to 105 °C to the following program: 1 cycle: 95 °C 5 min; 1 cycle: 65 °C 10 min; 1 cycle: 65 °C Hold; 60 cycles: 68 °C 1 min, 37 °C 3 s; 1 cycle: 68 °C Hold.
2. Preheat the thermal cycler to the first step of the program; 95 °C.
3. Add 5 μL of Blocker Mix to the 12 μL library pool solution. Seal the tube cap and vortex for 5 s. Spin the tube quickly to collect the liquid and release any bubbles.
4. Transfer the tube to a thermal cycler and incubate at 95 °C for 5 min, 65 °C for 10 min with a 65 °C indefinite hold following. In the meantime, prepare a 25% solution of SureSelect RNAse Block with nuclease-free water (0.5 μL SureSelectRNAse Block, 1.5 μL nuclease-free water, 2 μL total).
5. At room temperature, add 5 μL of Custom Probe and 6 μL of SureSelect Max Fast Hyb Buffer to the 2 μL 25% RNAse Block Solution. Mix thoroughly by vortex and spin down solution. Immediately return the custom probe to −80 °C storage to prevent degradation; do not keep it at room temperature for extended periods.
6. After 15 min, when the thermocycler reaches the 65 °C hold step, and without removing the library solution from the thermocycler, transfer 13 μL of the probe hybridization mix directly to the tube. Mix gently by pipetting up and down slowly 10 times. The total reaction volume should now be approximately 30 μL.
7. Seal the tube with a fresh cap. Vortex the sample and spin down briefly and quickly return to the sample to the thermocycler.
8. Skip the hold stage on the thermocycler to now perform 60 cycles of 68 °C for 1 min, 37 °C for 3 s. This stage should take approximately >45 min. In the meantime, preparation of the streptavidin beads should take place.

3.2.4 Prepare Streptavidin Beads and Buffers for Capture

1. Prepare six 0.2 mL tubes each containing 200 μL of Wash Buffer 2 for use in later capture wash steps. Keep the aliquots at room temperature for the time being.
2. Program and prewarm a second thermal cycler to a 68 °C hold step.
3. Vortex or shake vial of SureSelect Streptavidin beads to ensure complete resuspension, then aliquot 50 μL into a fresh 0.2 mL tube.
4. Add 200 μL of SureSelect Binding Buffer to the bead solution, mix thoroughly by pipetting up and down 20 times. Spin the solution down briefly.
5. Place the tube on DynaMag rack and allow a few minutes for the beads to settle. Carefully remove and discard the supernatant.
6. Repeat this washing step two more times, finally resuspending the washed beads in 200 μL of SureSelect Binding Buffer.
7. 10 min before the hybridization step ends, pre-warm the bead solution by incubating the tube in the second thermal cycler at 68 °C.

3.2.5 Capture the Hybridized Libraries

1. Once the hybridization program is complete, immediately add the hybridization solution to the pre-warmed beads in the thermocycler at 68 °C. Whilst keeping the tube on the thermocycler, mix thoroughly by pipetting the solution up and down 10 times. Seal the tube with a fresh cap.
2. Incubate the capture reaction tube for 10 min at 68 °C. In the meantime, transfer the previously prepared aliquots of SureSelect Wash Buffer 2 to a thermocycler set to 70 °C. Pre-warm the buffer until it is used for the washing steps.
3. Once the 10-min capture incubation is complete, briefly spin the tube to collect the liquid, then place the tube on the DynaMag rack. Allow the beads to settle for several minutes, then carefully remove and discard the supernatant.
4. Resuspend the beads in 200 μL of SureSelect Wash Buffer 1. Mix by pipetting up and down 15–20 times, until beads are fully resuspended.
5. Return the sample tube to the DynaMag, wait for the beads to settle and then remove and discard the supernatant. Make sure all Wash Buffer 1 is removed.
6. Resuspend the beads in 200 μL of SureSelect Wash Buffer 2 pre-warmed to 70 °C. To ensure complete resuspension, pipette the mixture slowly up and down 20 times while keeping the tube in the thermocycler set at 70 °C. Seal the tube with a fresh cap and vortex for 8 s. Briefly spin the tube down briefly to collect the solution and release any bubbles.

7. Incubate the sample solution in the thermocycler at 70 °C for 5 min. Upon completion, place the tube on the DynaMag at room temperature. Allow 1 min for the beads to settle on the magnet. Carefully remove and discard the supernatant.
8. Repeat this wash step five more times for a total of six washes.
9. Once the final wash is complete, briefly spin the tube to collect and remove any residual Wash Buffer 2.
10. Add 24 μL of nuclease-free water to the beads. Thoroughly resuspend by pipetting up and down 10 times. Keep the sample on ice until use in the PCR reaction.

3.2.6 Amplify the Captured Libraries

1. Preset a thermocycler with heated lid set to 105 °C to the following program: 1 cycle: 98 °C 45 s; 12 cycles: 98 °C 15 s, 60 °C 30 s, 72 °C 30 s; 1 cycle: 1 min 72 °C; 1 cycle: 4 °C Hold. Prewarm thermocycler before samples are loaded in. The number of cycles can be optimized based on input amount and desired yield.
2. Thaw and thoroughly mix the Amplification Master mix by gentle inversion (do not vortex) then briefly spin down.
3. Prepare a PCR reaction mix of 25 μL Amplification Master Mix and 1 μL SureSelect Post-Capture Primer Mix. Vortex briefly then place on ice.
4. Add the 26 μL of PCR reaction mix to the 24 μL capture bead solution. Mix thoroughly by pipetting up and down until the bead suspension is homogenous. Do not spin the samples at this step.
5. Place the sample tube in the thermocycler and run the program.
6. Once the program is complete, briefly spin the tube and place it on the DynaMag rack. Allow several minutes for the beads to settle on the magnet. Carefully remove and transfer the amplified library to a fresh tube and place on ice. Discard the remaining beads.
7. The library solution can be kept at 4 °C overnight or at −20 °C for prolonged storage.

3.2.7 Purify the Final Libraries Using Magnetic Purification Beads

1. Remove Agencourt AmpPure XP beads from the fridge and allow them to equilibrate to room temperature. Mix the bottle thoroughly to ensure homogenous resuspension.
2. Add 90 μL of Ampure XP beads to the 50 μL of amplified library. Mix thoroughly by pipetting up and down 20 times. Incubate at room temperature for 5 min to maximize recovery.
3. Place the tube on the DynaMag rack and allow beads to settle on the magnet for a few minutes. Carefully remove and discard the supernatant. It is recommended to leave a small amount of supernatant behind to avoid removing beads.

4. With the tube still on the DynaMag rack, add 180 μL of fresh 70% ethanol to the beads and incubate for 30 s at room temperature. Remove and discard the ethanol solution. Repeat this for two more washes.
5. Remove all ethanol solution from beads and remove tube from the rack. Allow the beads to air-dry for a minute to remove any residual ethanol solution. Be careful not to over dry the beads (beads appear cracked if overdried) as this will decrease elution efficiency.
6. Add 50 μL of nuclease-free water to the beads and mix thoroughly by pipetting up and down 20 times. Incubate at room temperature for 5 min.
7. Place the tube on the DynaMag rack and allow the beads settle on the magnet for a few minutes. Carefully remove and transfer the purified capture Hi-C library solution to a new 0.2 mL tube.
8. Quantify the concentration and size distribution of the library using a Qubit fluorometer and Agilent TapeStation, respectively. If amplification has been successful, a clear peak between 200 and 650 bp should be observed.
9. Store in fridge at 4 °C or −20 °C for prolonged storage.

3.2.8 Sequencing and Analysis of C-Hi-C Library

1. Submit the final Capture Hi-C libraries for pooled Illumina paired-end sequencing following to the NEBNext Ultra II DNA library kit recommendations (*see* **Note 16**).

4 Notes

1. Prepare NIB fresh, either on the day before or on the morning of the experiment. Add PMSF and β-mercaptoethanol immediately before nuclei isolation under a fume hood. PMSF has short half-life in water so should always be added just prior to use. As PMSF and β-mercaptoethanol are toxic, use appropriate safety equipment.
2. For our experiments, seeds are divided into 1.5 mL Eppendorf tubes and subjected to a 70% ethanol wash, followed by a 15-min 20% bleach sterilization and five washes in autoclaved sterile water.
3. The number of plates of *Arabidopsis thaliana* seedlings can be optimized according to experimental needs. This protocol is harvesting root organs, if studying whole seedlings or cotyledons, less tissue may be required.

4. Additional NIB can be used to rinse the nuclei suspension through the Miracloth and nylon mesh. Avoid squeezing the plant material.
5. Use wide-bore tips or trim the ends off 1000 mL pipette tips to avoid disrupting or damaging nuclei.
6. Intact nuclei should appear round/oval with clear defined edges. Number of nuclei should be counted using the following equation.

$$\text{Nuclei per } \mu\text{L volume} = \frac{\text{Counted nuclei}}{\text{Counted surface (mm}^2) \times \text{Chamber depth} \times \text{Dilution}}$$

7. Alternative restriction enzymes, such as HindIII, can be used in this protocol. If an alternative enzyme is chosen, it is important to consider its compatibility with the NEB r3.1 buffer and other steps in the protocol. The amount of DpnII enzyme added can also be optimized. For example, an additional 100 U can be added in the morning following the overnight treatment to improve digestion.
8. The ligation period can be optimized. If the sample is incubated overnight, an additional 100 U of T4 DNA Ligase can be added the next morning. The sample can then be incubated at 16 °C for an additional 5 h to ensure further proximity ligation.
9. qPCR can be used to assess the efficiency of digestion by the DpnII enzyme. A primer set that amplifies a fragment spanning a DpnII restriction site should be used, with pre-digestion and pre-ligation samples serving as controls. Digestion and ligation success can be assessed by performing a PCR using two forward primers flanking a DpnII restriction site. If digestion and ligation have been successful, a ligation-dependent PCR product is expected which can visualized by agarose gel electrophoresis.
10. The Bioruptor Pico recommends DNA concentrations of samples between 2 and 100 ng/μL. Smaller Bioruptor microtubes can be used; if so, dilute the Hi-C sample to smaller volume accordingly.
11. If DNA input is less than 100 ng, it is recommended to dilute the NEBNext Adaptor. For example, if DNA input is between 5 and 100 ng, the adapter should be diluted ten-fold. The appropriate level of dilution can be optimized experimentally.
12. At this stage, the effectiveness of the bead capture can be assessed by performing a PCR to detect a ligation-dependent PCR product. Using a small aliquot of the bead-bound Hi-C sample (1 μL) and two forward primers that flank a DpnII restriction site. If the ligation-dependent product can be formed, this indicates that the Hi-C sample has been successfully pulled down by beads.

13. It is important to keep a record of which NEBNext Index i7 Primer is used as these indexes are essential for demultiplexing pooled libraries during sequencing. If the same library requires amplification multiple times, use the same i7 index to avoid confusion. Do not concentrate the beads in an attempt to fit the entire sample into a single PCR reaction, as this can result in poor amplification efficiency.
14. Do not discard the remaining beads. Resuspend them in 25 μL 0.1 × TE buffer. These can be kept at 4 °C or −20 °C to be potentially used again for library amplification. If reusing beads, wash several times with 1xB&W buffer before use. Always use the same Index i7 primer as previous reaction.
15. For our CHi-C project: using the Col-CEN assembly, a 250 kb genomic region encompassing the thalianol cluster was selected. Probes were designed to cover all DpnII restriction fragments within this region, resulting in a total of 870 unique probes designed for the target region.
16. Sequencing depth should be chosen to ensure sufficient coverage to achieve high-resolution contact maps for the captured regions. It is important to note that Capture Hi-C introduces specific biases therefore normal Hi-C pipelines may not be appropriate. For accurate interpretation, we recommended using a combination of HiCUP and CHiCANE pipelines [13, 14].

Acknowledgments

This work was supported by Royal Society awards RF\ERE\221026 and URF/R/221001 to H.W.N and the University of Exeter (G.J. and H.W.N.).

References

1. Dekker J, Rippe K, Dekker M, Kleckner N (2002) Capturing chromosome conformation. Science 295(5558):1306–1311. https://doi.org/10.1126/science.1067799
2. Zheng H, Xie W (2019) The role of 3D genome organization in development and cell differentiation. Nat Rev Mol Cell Biol 20(9): 535–550. https://doi.org/10.1038/s41580-019-0132-4
3. Pei L, Li G, Lindsey K, Zhang X, Wang M (2021) Plant 3D genomics: the exploration and application of chromatin organization. New Phytol 230(5):1772–1786. https://doi.org/10.1111/nph.17262
4. Han J, Zhang Z, Wang K (2018) 3C and 3C-based techniques: the powerful tools for spatial genome organization deciphering. Mol Cytogenet 11(1):21. https://doi.org/10.1186/s13039-018-0368-2
5. Lieberman-Aiden E, van Berkum NL, Williams L, Imakaev M, Ragoczy T, Telling A et al (2009) Comprehensive mapping of long-range interactions reveals folding principles of the human genome. Science 326(5950): 289–293. https://doi.org/10.1126/science.1181369
6. Feng S, Cokus SJ, Schubert V, Zhai J, Pellegrini M, Jacobsen SE (2014) Genome-

wide Hi-C analyses in wild-type and mutants reveal high-resolution chromatin interactions in *Arabidopsis*. Mol Cell 55(5):694–707. https://doi.org/10.1016/j.molcel.2014.07.008

7. Dong Q, Li N, Li X, Yuan Z, Xie D, Wang X et al (2018) Genome-wide Hi-C analysis reveals extensive hierarchical chromatin interactions in rice. Plant J 94(6):1141–1156. https://doi.org/10.1111/tpj.13925
8. Concia L, Veluchamy A, Ramirez-Prado JS, Martin-Ramirez A, Huang Y, Perez M et al (2020) Wheat chromatin architecture is organized in genome territories and transcription factories. Genome Biol 21(1):104. https://doi.org/10.1186/s13059-020-01998-1
9. Dryden NH, Broome LR, Dudbridge F, Johnson N, Orr N, Schoenfelder S et al (2014) Unbiased analysis of potential targets of breast cancer susceptibility loci by Capture Hi-C. Genome Res 24(11):1854–1868. https://doi.org/10.1101/gr.175034.114
10. Huang AC, Jiang T, Liu Y-X, Bai Y-C, Reed J, Qu B et al (2019) A specialized metabolic network selectively modulates *Arabidopsis* root microbiota. Science 364(6440). https://doi.org/10.1126/science.aau6389
11. Nützmann H-W, Doerr D, Ramírez-Colmenero A, Sotelo-Fonseca JE, Wegel E, Di Stefano M et al (2020) Active and repressed biosynthetic gene clusters have spatially distinct chromosome states. Proc Natl Acad Sci 117(24):13800–13809. https://doi.org/10.1073/pnas.1920474117
12. Jervis, G. (2026) https://BioRender.com/u08lqg3. Created in BioRender.
13. Wingett S, Ewels P, Furlan-Magaril M, Nagano T, Schoenfelder S, Fraser P et al (2015) HiCUP: pipeline for mapping and processing Hi-C data. F1000Res 4:1310. https://doi.org/10.12688/f1000research.7334.1
14. Holgersen EM, Gillespie A, Leavy OC, Baxter JS, Zvereva A, Muirhead G et al (2021) Identifying high-confidence capture Hi-C interactions using CHiCANE. Nat Protoc 16(4):2257–2285. https://doi.org/10.1038/s41596-021-00498-1

INDEX

Hans-Wilhelm Nützmann and Ramesha Thimmappa (eds.), *Metabolic Gene Clusters: Methods and Protocols*, Methods in Molecular Biology, vol. 3054, https://doi.org/10.1007/978-1-0716-5384-5,

S

T

U

V

W

Y

MIX
Papier aus verantwortungsvollen Quellen
Paper from responsible sources
FSC® C105338

If you have any concerns about our products,
you can contact us on
ProductSafety@springernature.com

In case Publisher is established outside the EU,
the EU authorized representative is:
Springer Nature Customer Service Center GmbH
Europaplatz 3, 69115 Heidelberg, Germany

Printed by Libri Plureos GmbH
in Hamburg, Germany